Praise for Richard M. Restak

RECEPTORS

"A noted neurologist and author of *The Brain* and *The Mind* . . . Restak, who has long listened to Prozac and other mind-altering drugs, provides a lucid, balanced and helpful history of the steps leading us to this new frontier."
—*Publishers Weekly*

"Citing a series of discoveries showing that receptors are the gatekeepers of everything we know and feel, Restak makes the important point that everybody's brain is always on drugs—whether those drugs are the body's own endorphins, prescription antidepressants or street-synthesized LSD. . . . [He] takes the tried and true approach of making a series of points by telling a series of stories, most of them focused around a particular scientist in history. . . . The historical tales make good reading. Through them we learn that as scientists and psychiatrists saw the specific influences of drugs on mood and behavior they also came to recognize that mental illness is not a supernatural malady of the psyche but a biological phenomenon potentially treatable with medicines."
—*The Washington Post Book World*

THE MIND

"A neurologist with several books to his credit, Dr. Restak is an accomplished science writer with a gift for making complex scientific subjects accessible to the general reader."
—*The New York Times Book Review*

"A wonderful book . . . the subject matter is inherently fascinating and Restak is a whiz at narration."
—*The Washington Post*

"Immensely informative, crammed with challenging ideas . . . [and] as enthralling as the elegant mind which composed it."
—Rod MacLeish, National Public Radio

THE BRAIN

"A very special achievement. I can't think of anything on the subject that combines authentic knowledge with literary craftsmanship as beautifully as this book does."
—Norman Cousins

"Rich in imagination . . . intriguing and thoroughly understandable."
—*The New York Times Book Review*

"Excellent popular science: It's fun to read, high in information content, scientifically honest, and should have broad appeal."
—*Los Angeles Times*

BOOKS BY RICHARD M. RESTAK, M.D.

RECEPTORS

Richard M. Restak, M.D.

BANTAM BOOKS

NEW YORK TORONTO LONDON SYDNEY AUCKLAND

RECEPTORS
A Bantam Book
PUBLISHING HISTORY
Bantam hardcover edition / April 1994
Bantam trade paperback edition / April 1995

Grateful acknowledgment is made for permission to reprint from
"The Discovery of LSD and Subsequent Investigations on Naturally
Occurring Hallucinogens" by Albert Hofmann published in
Discoveries in Biological Psychiatry, ed. by Frank Ayd and
Barry Blackwell, copyright 1984 by Ayd Medical Communications.

An excerpt from this book was published in somewhat
different form in the magazine *The Sciences.*

Bantam Books are published by Bantam Books, a division of Bantam Doubleday Dell
Publishing Group, Inc. Its trademark, consisting of the words "Bantam Books" and
the portrayal of a rooster, is Registered in U.S. Patent and Trademark Office and
in other countries. Marca Registrada. Bantam Books, 1540 Broadway, New York,
New York 10036.

PRINTED IN THE UNITED STATES OF AMERICA

BVG 0 9 8 7 6 5 4 3 2 1

TO MY DAUGHTER ANN

ACKNOWLEDGMENTS

*T*his book would not have been possible without the assistance and encouragement of the following, who graciously made time for interviews, responded to inquiries, or suggested resources: Joel Elkes, M.D.; Alfred Freedman, M.D.; Miles Herkenham, Ph.D.; Conan Kornetsky, Ph.D.; John W. Olney, M.D.; Oakley S. Ray, Ph.D.; Solomon Snyder, M.D.; Lars Terenius, Ph.D. I owe special thanks to Kenneth Kellar, Ph.D., who read the entire manuscript and made many helpful suggestions. Thanks are due to my editor, Ann Harris, to my agent and friend Sterling Lord, and to my family: my wife Carolyn, and my daughters, Jennifer, Alison, and Ann.

CONTENTS

PART I

CHAPTER 1 Disordered Thought, Disordered Molecule   3

CHAPTER 2 The Brain   12

CHAPTER 3 Magic Intoxicants   33

CHAPTER 4 Dr. Hofmann and the Magic Circle   42

CHAPTER 5 Lethargy in a Guinea Pig   61

CHAPTER 6 Clasping the Neuron   68

CHAPTER 7 The Anatomy of Melancholy   74

PART II

CHAPTER 8 Russian Dolls   89

CHAPTER 9 Double Agents   103

CHAPTER 10 Good Drugs—Bad Drugs   116

CHAPTER 11 The Dust of Angels   126

CHAPTER 12 Dr. Freud and the Devil's Bargain   135

CHAPTER 13 Mother's Little Helper   155

CHAPTER 14 Mr. Guinea Pig   168

PART III

CHAPTER 15 Tracking the Marijuana Receptor   183

CHAPTER 16 Designer Brain   202

Further Readings   217

Index   219

PART I

CHAPTER 1
DISORDERED THOUGHT, DISORDERED MOLECULE

At a Fourth of July party I recently attended, the hostess proposed a game. She provided all her guests with name tags and affixed them to each of their foreheads, but in place of their own names she had put the name of a historical person, place, or event. People could read the names on the others' brows but couldn't see and therefore didn't know their own. The object of the game was for each person to discover his or her identity by asking questions of the other guests.

That night I had a dream in which I was at a similar gathering, but the guests and theme of this dream party were different. All the world's most distinguished brain scientists were there, and affixed to their foreheads were the names of parts of the brain. Some of the labels referred to large brain areas such as the temporal lobe or to microscopic nerve cells like the neuron; others identified chemical messengers—neurotransmitters—within the brain; still others identified processes like synaptic transmission, whereby a chemical message is transferred across the synapse, the tiny space that separates nerve cells. Since I have been studying the brain all of my adult life, none of this seemed surprising to me: not the setting, nor the participants, nor the game itself.

In my dream I am talking to a neuroscientist who has the neurotransmitter *dopamine* written on her brow. Thanks to her clever questioning, she has already discovered her own identity and is trying to help me discover mine. But whatever I ask her, no matter how skillfully she answers, I cannot seem to guess what is written on my forehead.

Finally, the woman reaches out, puts her arms around me, and plants a lovely kiss on my lips. She then steps back, her eyes wide open, and asks, "Do you know now?" And of course, I do. I am the *receptor* for dopamine. Like two lovers, dopamine and its receptor, like all brain neurotransmitters and their receptors, embrace across the synaptic junction.

My dream intrigues me for a couple of reasons. One is the Fellini-like situation of brain scientists using their own brains in order to guess the name of the brain element they have been assigned to represent. Further, all the participants in this fantastic dream situation, when taken together so that they include all of the various brain parts and processes, represent the brain in its entirety. Moreover, on the basis of their assignment of a single brain element each, the players are being challenged to comprehend not only their individual identity and their participation in the brain's operation but the operation of the brain as a whole. In essence, parts of the brain are being challenged to understand the functioning of the whole.

Is this possible? Can the study of one component—say, the most elemental unit, the neuron—provide insights into the functioning of the whole? On the heels of that consideration comes the image of my seductive interrogator, the brain messenger reaching out and embracing the specific receptor—in this exciting dream, me. A neuron sends a message to another neuron by releasing a neurotransmitter across the synapse to a receptor site on the membrane of the target neuron. That receptor is specialized to receive that neurotransmitter. This is how nerve cells of the brain communicate, both within the brain itself and outside it, with the entire body. What part does this intricate neurotransmitter-receptor interaction play in the operations of the brain?

In this book, I propose that our understanding of these operations is being remarkably advanced by what we are learning about the relationship of the brain messengers to their receptors. This burgeoning understanding has far more than theoretical importance. In the next decade receptor research promises to bring about remarkable, wide-ranging advances in our understanding of human behavior and in treatments for thus-far-incurable mental illnesses and for drug addiction. The application of these advances extends beyond understanding extraordinary mental states. It also holds promise of improving normal functions: enriching memory, enhancing intelligence, heightening concentration, and altering for the good people's internal moods.

Throughout history, insights into how the brain works have been shaped by the technology and the scientific biases of a particular time. The early Greeks, influenced by the technology of aqueducts, thought of mental processes in terms of the flow of bodily fluids. In the seventeenth century the philosopher Descartes compared brain functions to the operations of machines. The nineteenth century emphasized anatomy, the physical connections of one brain part to another. This view of the brain was stimulated by widespread fascination with railway lines and the demonstration that communication depends upon physical connectivity over distance. A similar connectionist view was espoused in the early years of the twentieth century by a famous brain scientist, Sir Charles Sherrington, who compared brain operation to a telephone switchboard.

Today brain function is often compared to circuits and computations, reflecting the influence of electronics and computers in our age. But none of these technological metaphors ever hit the mark with sufficient precision to withstand the assault of the next innovation. Analogies to other processes and "machines," no matter how sophisticated, falter when placed against the complexity and uniqueness of the human brain.

If the brain is not like an aqueduct or a railroad or a telephone switchboard or a computer, and if future technological metaphors are just as likely to be displaced, what then is the best way of thinking about it?

Over the past quarter-century, and especially the last ten years, a new way of thinking about the brain and the mind—functionally defined as operations carried out by the brain—has flourished. Rather than emphasizing anatomy, which is essentially another mechanical theory of how the brain works, this new theory concentrates on understanding the brain on the chemical and molecular level. Since this can be a somewhat difficult point to grasp, at least initially, an incident that occurred during my residency may help illuminate it.

In my training years in neurology I worked under a morose, short-tempered teacher who was far more interested in the chemistry of mental and neurological illnesses than in the care of patients afflicted with these illnesses. He was not an uncaring man; rather, he had become so immersed in his research that what he could see or hear or touch failed to stir within him the passion that the invisible molecular world inspired in him.

One afternoon I walked into his lab unannounced to ask for his help in managing a patient I had just seen. I was just starting my residency training, and I was often uncertain about how to proceed with the patients who sought care in this big-city emergency room.

As I spoke, my chief listened to what I was telling him, but he continued to stare down at the chemical formulas on the papers spread out before him. At the conclusion of my presentation, he gave me an incisive, accurate diagnosis, and even more welcome to my ears, he told me what to do to help my patient.

Then as I was turning to leave, he said something that has taken me twenty years to understand: "Now, please go away and tend to that macromolecule in the emergency room, and leave me here to deal with these micromolecules."

Over the next several days we neurology residents would break into bursts of laughter whenever we thought about his strange remark, this weird idea of a living, breathing human being as a "macromolecule" that could be compared to or equated with molecules in the brain. Two decades later, my teacher's comment seems much less bizarre. Research on the human brain suggests that we share certain properties and reactions with the micromolecules of which we are constituted. Perhaps the one who has come closest to understanding this continuum from the very small to the very large was the man who suggested, "Behind every crooked thought there is a crooked molecule." In short, all things mental—both normal functions and disorders of thought and emotion—originate from some corresponding order or disorder at the molecular level. Over untold millennia nature has evolved certain principles and, since they work, employs them again and again at various levels of the organism. Further—and this is the theme that we will be developing in this book—nature displays a marvelous parsimony, in which events at one level mirror what is going on several higher or lower orders away.

The common unit at all levels is information. Information can take the form of the words on a page, or the signal that instantaneously moves a finger away from a hot surface, or the swirl of neurotransmitters within a synapse between neurons. Indeed, as Umberto Eco pointed out in a conversation several years ago, the whole world and everything in it, including the brains of everyone in the world, can be thought of in terms of information. There is information in the shape of a leaf, in the sound of a sparrow, in the molecular structure of water, and in the multitude of chemicals that make up our bodies. Over the aeons the human brain has evolved to encode and decode myriad forms of information. It does this through neurotransmitters, each of which has its own chemical formula but shares a common purpose with the others: to communicate information from one brain cell to another.

The recognition that communication at the molecular level may be at the root of brain functioning had its impetus in the field that is now called neuropsychiatry: a science rooted in the belief that mental illness and emotional disturbances can be understood and treated in terms of disturbances in the functioning of the brain.

Neuropsychiatry is a comparatively new field, although mental illness is as ancient as human history. Only fifty years ago, theories about the origin and management of mental illness still depended upon unproved and essentially unprovable assumptions about human behavior. Although psychoanalysis and other "talking" therapies had their place in the treatment of neuroses and milder forms of mental illness, they were generally far less useful in treating psychoses. The number of people confined to mental hospitals in the late 1940s was at an all-time high, most of them suffering from major psychotic disturbances, but for the most part psychiatrists operated outside the "medical model" in treating them. The prevalent treatment methods consisted of efforts to control a patient's disturbed behavior—restraints, cold packs, isolation cells—and the most pervasive of all, neglect. The only drugs then available had a similar purpose: control. But they were unreliable and often dangerous. Barbiturates brought about some calming of agitated patients, but at the risk of oversedation and dependence. Bromides enjoyed some popularity for a time, but they often induced their own form of madness, known as bromidism. Biological treatments, also called somatic treatments, were limited to electric shock and insulin coma therapy. Not only did they fail to bring about reliable and repeatable improvements in patients, they were also frightening and were perceived as dangerous and repugnant. Insulin coma therapy deliberately reduced a patient's blood sugar until coma set in; the risks included brain damage and death. Electric shock treatment, also called electroconvulsive therapy or ECT (which is used today with far greater safety and effectiveness), was often applied inappropriately and without adequate precautions to protect patients from injury.

Though trained as physicians, most psychiatrists of the time had little interest in neurology, the science of the brain and the nervous system. By current standards that may sound like a serious indictment, but it is not meant to be. This was still an era in which mental illness was generally perceived—by almost all doctors, not only by psychiatrists—as resulting from disturbances in the mind, as distinct from the body. Psychiatrists actually had more in common with psychologists, philosophers, and others who looked upon human personality in ways that excluded biology, than they did with other physicians.

There were, however, a small number of psychiatrists who recognized the significance of the brain in mental dysfunction. Some had been formally trained in the brain sciences—neuropsychiatrists who shifted their professional interests from neurology to psychiatry. (This was the path Freud had taken many years before.) But most lacked professional training in neurology, and had only a gut feeling that the mind could not be properly understood without reference to the brain and nervous system. They had no way of proving this, however.

Today, all that is changed. Our understanding of the brain and central nervous system has taken a quantum leap, and with it has come the development of mind-altering drugs that literally act on the "crooked"—disordered—molecules that have been found to characterize many mental disorders. Today we know that our thoughts and emotions, both normal and disturbed, are the result of chemical processes going on within the brain. Alter these processes, and the thoughts and emotions will be altered. The new drugs that change the chemistry of the brain make clear that we can change our internal states deliberately.

Chemical alteration of the brain is not something new in history; human beings have done it for millennia. But our increasing understanding of the processes involved is very new. For thousands of years people have sought altered states of consciousness by distilling mind-altering ingredients from plants. Some of these are exotic, but many are so commonplace that we do not even think of them as drugs. Few people complete a day without recourse to a stimulant—coffee or tea or chocolate. Is there anyone who has never taken an aspirin for a headache? When we want to relax, some of us reach for a drink or a cigarette. These are all brain- and mind-altering substances, and since they are all legal, there has usually been no shame or guilt or embarrassment about using most of them.

That has recently changed in some respects. It is much harder now than it was five years ago to light up a cigarette at a cocktail party without being challenged on the dangers of passive smoking. Overall, the consumption of alcohol too is down, particularly the hard liquors that were once so popular. Part of the reason for these changes in attitude stems from our increasingly sophisticated knowledge about addiction. We now recognize alcohol and tobacco as substances capable of rendering us unable to live without them.

At a different level of urgency is our recognition of the consequences of widespread cocaine, heroin, and marijuana use. We see these consequences around us everywhere: deaths from drug overdoses, the murders of drug dealers by one another, the growing violence among the young, the dangers epidemic in our cities. Drugs are frequently associated with automobile accidents, shipwrecks, plane crashes, and train disasters; people today commonly embark on a journey on any form of transportation with a gnawing fear that the operator might be on drugs. Drug addiction and the apparently intractable problem of what can be done about it are a constant presence in our lives.

One thing is certain: Blanket prohibition of mind-altering substances won't work; nor will efforts to differentiate between "good" and "bad" drugs, for reasons we will explore in Chapter 10. Brain and mind alteration through chemistry is a fact of our lives. Illicit drugs apart, try to imagine what our world would be like if suddenly, overnight, all pain-killers, sedatives, caffeine-containing compounds, nicotine, and everything containing alcohol; all tranquilizers, antidepressants, and anticonvulsants were simply to disappear. Is there anyone so harsh and puritanical among us that he or she would deny pain relief to cancer patients, or to those suffering the chronic pain of arthritis or migraines? Who would refuse to relieve, if not "cure," mental patients of their delusions and anguish with appropriate tranquilizers? Do not tea and coffee, when taken in sensible amounts, make life more pleasurable? And would we really be happy in a world without chocolate? Although some of us might prefer a world without alcohol—a world that has never existed, even in earliest times— the vast majority would not like to do away with this mind-altering substance altogether. Rather than eliminate all these substances— an impossibility—we would like to know how to control them better. In order to do this we must understand how they affect our brains.

All such substances, licit and illicit, therapeutic and commonplace, natural and synthesized, provide windows into the functioning of the brain. What is now emerging from the laboratories of molecular biologists and psychopharmacologists are new brain- and mind-altering drugs that have an even vaster potential to alter how we feel and act. To understand these explosive developments, let us turn to the organ upon which these substances act, especially those areas that play important roles in emotions and behavior.

It is time to look at the brain.

CHAPTER 2

THE BRAIN

On the macroscopic level the brain lends itself to partitioning according to time-honored systems based largely on its anatomy. But the map is not the territory: *All* proposed divisions within the brain are highly artificial and are based on our need to separate things into neat, easily understandable units. We must always remember that the brain functions as a whole. Bearing that caveat in mind, figure 1 shows us a useful schema that we can refer to when we are envisioning how our brains are organized, beginning with the outermost surface, which is responsible for our highest conceptual and perceptual functions, and working downward. The emphasis here is on those parts of the brain that play important roles in emotions and behavior.

The cerebrum, the largest part of the brain, is divided into two hemispheres that consist of an inner core of white matter and a deeply convoluted outer layer, the gray matter called the cortex. The hemispheres are specialized for the processing of specific information. For convenience, we separate them into four areas: the occipital, temporal, parietal, and frontal lobes. Vision is processed in the occipital lobes, toward the rear. The temporal lobes, at the sides, are important for hearing, some aspects of memory, and a person's sense of self and time. In the parietal lobes, along the top, is the sensory cortex; here touch, pressure, and other skin-mediated sensations are processed. The motor cortex, which processes willed as opposed to reflex movements, lies immediately before the central sulcus, which roughly divides the brain into anterior (front) and posterior (rear) portions. In the very front, just behind the forehead, are the frontal lobes, the largest lobes of all. The most forward portion, the prefrontal area, enables us to think: to foresee the consequences of our actions, to plan for the future, and to work out philosophical, religious, and ethical systems—all those activities that differentiate us from the other creatures on earth.

Though the two hemispheres of the cerebrum look very much alike, each controls the opposite side of the body, and each processes information in different ways. This has given rise to the concept of the "two brains," in which the right hemisphere operates holistically, performing visual spatial tasks, and the left excels at breaking things down into their component parts, specialized for language and logic. Nonetheless, it must be remembered that many functions involve both hemispheres, which are in constant communication with each other across the corpus callosum, the thick bridge of nerve fibers that joins them.

Below the two cerebral hemispheres, hidden from external view, are the basal ganglia. These structures, which answer to such exotic names as the globus pallidus, the putamen, and the caudate, are responsible for coordinating and initiating all body movement, including—in combination with the cerebellum—unconscious, automatic elements of movement. When you stand in line at a supermarket, these areas help you maintain the necessary tone so that you neither flop down like a rag doll nor stiffen to the point that you can't move at all.

The limbic system, also called the "primitive" or old brain,

MAJOR AREAS OF THE BRAIN

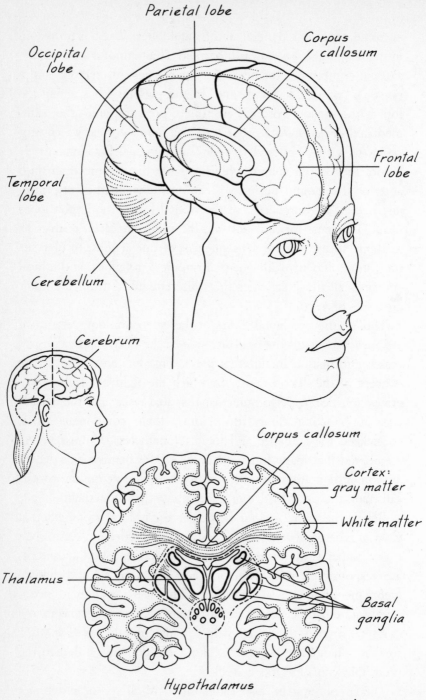

Parietal lobe

Corpus callosum

Occipital lobe

Frontal lobe

Temporal lobe

Cerebellum

Cerebrum

Corpus callosum

Cortex: gray matter

White matter

Thalamus

Basal ganglia

Hypothalamus

FRONTAL VIEW (Cross section through cerebrum)

IMPORTANT STRUCTURES BENEATH THE CEREBRAL HEMISPHERES

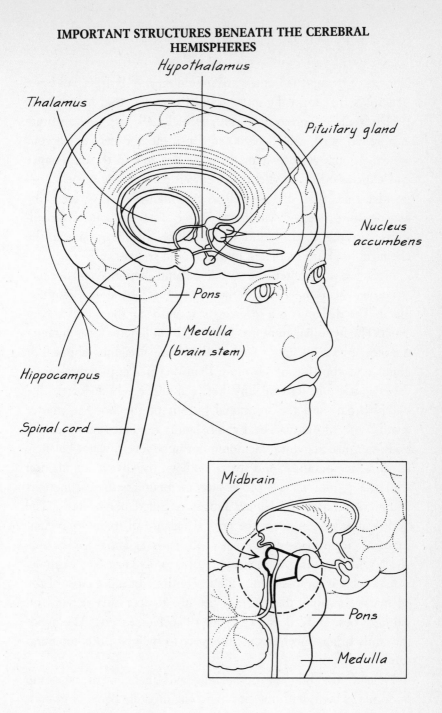

Hypothalamus

Thalamus

Pituitary gland

Nucleus accumbens

Pons

Medulla (brain stem)

Hippocampus

Spinal cord

Midbrain

Pons

Medulla

consists of a ring (the Latin word *limbus* means "border") of brain structures that form the inner wall of each cerebral hemisphere. It shares direct connections with virtually all parts of the brain and spinal cord, which together constitute the central nervous system. Its importance stems from its key role in the experience and expression of emotion. Indeed, all emotions and feelings—positive ones like love as well as potentially negative ones like anger—are processed within the limbic system. Many of the advances in the control and manipulation of the brain at the chemical level are mediated by the limbic system.

Next we encounter the thalamus and hypothalamus, small structures near the base of the brain, just above the pituitary gland. The thalamus is a receiving station for all the senses except smell. The hypothalamus, just below it, is millimeter for millimeter the most powerful subdivision of the brain. Weighing about four grams and constituting no more than one percent of total brain volume, it is the critical link between the cerebral cortex, the limbic brain, and the hormonal output of the body's "master gland," the pituitary. The hypothalamus regulates many of the body's critical activities, including hormone levels, water balance, appetite, temperature, and sexual activity, usually in conjunction with the pituitary. The process of maintaining these functions within very narrowly defined ranges is called homeostasis. The body's ability to preserve the critical balance of its internal environment, regardless of external changes, is essential for its survival. Wide fluctuations in temperature, to take just one example, would make it impossible for life to continue because most of the chemical reactions that sustain life are temperature-dependent, working best at 37 degrees Celsius (98.6 Fahrenheit). The hypothalamus brings mechanisms into play to maintain this temperature.

Beneath the limbic-hypothalamic and thalamic areas is the midbrain, which, with the pons and the medulla beneath it, constitutes the brain stem. In the brain stem are centers that control blood pressure, temperature, heart rate, and breathing. Since these activities take place outside of our awareness, the areas responsible for them are called the autonomic nervous system. Many mind- and brain-altering drugs exert powerful effects on the autonomic nervous system. For example, some tranquilizers cannot be used in patients who have certain conditions because they will lower their blood pressure to dangerous levels. Other drugs speed up or slow down the heart to rates that could pose a danger to life in persons sensitive to them.

There are two branches of the autonomic nervous system: the sympathetic and the parasympathetic branches. In most instances these act in opposition to each other. In emotion-arousing circumstances the sympathetic branch prepares for stress or danger by initiating the "flight or fight" response: a rise in blood pressure, an increase in heart rate, an acceleration in breathing, and other bodily responses directed toward self-preservation.

Stimulation of the parasympathetic branch brings on the "relaxation response," in which the heart rate, blood pressure, and respiration decrease, leading to a feeling of repose and well-being. Meditation and various forms of stress management aim at inhibiting the sympathetic response and heightening the parasympathetic response.

Drugs can be employed to stimulate either the sympathetic or the parasympathetic response. Amphetamine (about which we will say more in Chapter 12) possesses a chemical structure similar to norepinephrine, a natural neurotransmitter of the sympathetic branch, and therefore provokes the same response as norepinephrine. Drugs like amphetamine are therefore referred to as sympathomimetic—that is, they mimic the action of the sympathetic branch. Other drugs induce a state of relaxation by influencing the parasympathetic branch directly or by decreasing sympathetic influence.

Indeed, one can think of the sympathetic and parasympathetic branches of the autonomic nervous system as balancing each other, like two sides of a scale. Removing weight on one side results in an overbalance on the other. If a public speaker suffering from performance anxiety, for example, is given a blocking drug that prevents norepinephrine from attaching to its receptors, the influence of the sympathetic branch will be diminished and the action of the parasympathetic branch will be enhanced. Our speaker will perform without anxiety.

As human beings, we understandably focus on those capacities of the brain that are unique to our species: perception, memory, imagination, language, and creativity. But the fact is that the brain's elaborate structure and interconnections evolved not so much to develop these "higher" functions as to regulate the intricate processes of the body that make life possible. Once again, communication is the key, communication among the billions and billions of neurons that constitute the brain, receiving information from other nerve cells within it and throughout the body, processing it, and sending messages that affect every organ and cell.

The functional unit of the brain is the neuron, a cell that differs from other body cells in that it is specialized for the function of information processing. It has four components: the cell body or soma, the axon, the dendrite, and the synapse. The cell body manufactures all the proteins and other chemical substances that the neuron needs to perform its functions and to sustain itself. The nucleus in the cell body, like the nuclei in all other cells, contains an individual's entire genetic endowment of DNA.

The second component of the neuron, the axon, is a delicate tubular extension from the cell body. It is the pathway along which nerve impulses are *sent* from the neuron to the synapse. Nerve impulses flow from the neuron along the axon to the synaptic terminal. When neurons are linked to many target neurons, forming a network, their axons have branches. Axons can vary enormously in length, from only a few millimeters to more than a meter for those neurons that activate the muscles in our arms or legs.

Dendrites, in contrast to axons, are multiple tubular extensions of the cell body. They are the *receivers* of neuronal impulses from other neurons. Some dendrites are richly branched, with multiple receptor sites that enable them to make contact with many other neurons. Other dendrites have only a few receptor sites.

As noted earlier, communication between neurons takes place across the minute space called the synapse, or synaptic cleft. An outpouching of the terminal aspect of the axon, called a bouton (or button), projects into this space. Contained within the bouton are vesicles—spherical structures responsible for activating and releasing neurotransmitters into the synaptic cleft.

Like all living cells, the neuron is bound by a cell membrane that separates it from the rest of the world. It is this nerve cell membrane, within which is a concentrated aqueous solution of chemicals and other structures, that makes communication between neurons possible. This is because signaling within the brain—in essence, the flow of information from one nerve cell to another—involves the passage of ions, electrically charged chemical particles, through separate, minute channels in the cell membrane. The nerve cell exists in a state of creative tension because of the concentrations of various ions across its membrane. Variations in these concentrations generate the nerve impulse, or *action potential,* which allows for the transmission of information between neurons.

Four ions are most important in this process: sodium, potassium, calcium, and chloride. Differences in the permeability of the cell membrane to two of these ions (a high permeability to potassium, a much lower permeability to sodium) create a resting electrical potential across the cell membrane.

Each ion passes in or out of the neuron via its own channel, which is built into the cell membrane. Also built into the membrane is an ion pump that secretes two sodium ions for every potassium ion that is pumped into the cell. The resulting uneven distribution of ions—high intracellular potassium and low extracellular potassium, coupled with high extracellular sodium and low intracellular sodium—produces a voltage difference between the outside and the inside of the nerve cell. This voltage difference

NEURON

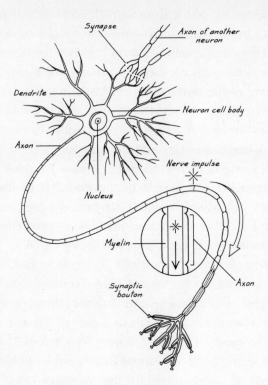

SYNAPSE

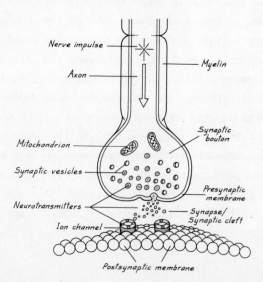

confers on the neuron its capacity to generate and transmit the
wave of electrochemical depolarization that *is* the nerve impulse.

Physically, the channels for these four ions are complex protein
molecules. These molecules can exist in a variety of different
states—think of them as "gates" that can be either *closed,* so that
no ions can pass through, or, with a change in the molecular
configuration of the protein, *open* so that particular ions are free
to move through.

In order to generate a nerve impulse, a cell membrane must
first be depolarized. Depolarization begins when a sodium channel
across the membrane is briefly opened and sodium ions pass
through it into the cell. This results in a reduction of voltage.
When the voltage reduction reaches a critical level—about -35
millivolts—adjacent sodium channels are also opened, generating
an action potential. This nerve impulse spreads down the axon,
as the voltage-controlled sodium channels open sequentially like
falling dominoes.

Once the action potential reaches the end of the axon, the
electrical transfer of information ceases and messages are conveyed by chemical means alone: neurotransmitters. At the axon
terminus of the sending neuron, the action potential triggers the
presynaptic membrane to release a neurotransmitter into the synaptic cleft. It diffuses across the synaptic cleft and binds to its
receptor on the postsynaptic membrane on the dendrite of the
target cell, opening or closing channels on that membrane and
thereby initiating a current flow. Depending on the type of channel that the transmitter has gated chemically, the action potential
exerts either an excitatory effect on the target neuron, influencing
it to "fire," or generate an electrical impulse of its own, or an
inhibitory effect, preventing it from "firing." Neurons are continuously adding up excitatory and inhibitory inputs that impinge
on them, in order to determine whether an action potential will
be generated.

Neurons may be in different states of preparedness for firing.
Some may be just a few millivolts away from the critical voltage

level, requiring only a minimum of excitation in order to reach this "threshold," while others may have a preponderance of inhibitory influences that must be overcome in order for the nerve impulse to be generated and the neuron to fire.

✜   ✜   ✜

The concept of a transmitter and its receptor is not new. In 1900 the German biological chemist Paul Ehrlich introduced the idea of receptors to explain the action of poisons: "chemical substances are only able to exercise an action on the tissue elements with which they are able to establish an intimate chemical relationship. . . . This relationship must be specific. The chemical groups must be adapted to one another . . . as lock and key." The relationship of a neurotransmitter to its receptor is actually much more dynamic and mysterious than a lock and key, but it is still a useful analogy.

Neurotransmitters, like all chemicals, can be described by their structures—the ways in which their component atoms come together to form molecules. Each neurotransmitter molecule has a unique three-dimensional structure that differentiates it from all others. The receptor molecule on the target neuron also has a specific structure, and it is into its "lock" that the specific neurotransmitter "key" will fit. Nonetheless, a given neurotransmitter may find itself in competition for a receptor with another neurotransmitter of slightly different configuration. The receptor will adhere to only one of them. Thus, depending on its degree of fit with the receptor, the neurotransmitter may be taken up by the receptor in one instance but may fail to activate it in another one, when an additional molecule is present that fits the receptor even better. As in dating and mating, what is finally settled for is always a function of what is available. Since neurotransmitters and receptors vary widely in their affinity for each other, depending on what other neurotransmitters are available, they are significantly different from the metaphor of a key and lock.

✛   ✛   ✛

Neurotransmitters are of four main chemical types. The most prevalent are simple amino acids, which are involved in rapid point-to-point communication between neurons. These are entirely natural substances and are found throughout the brain; they are also normal constituents of our diets, which puts a new spin on the old adage that you are what you eat. Glycine, glutamate, and aspartate are three of the most important of the twenty common amino acids that function as neurotransmitters. Glutamate is the primary excitatory messenger within the brain. Gamma-amino butyric acid (GABA), the most prevalent inhibitory neurotransmitter, is made from glutamate with the help of a single enzyme. (Enzymes—often referred to as nature's catalysts—are chemicals that speed up the splitting apart or stitching together of molecules but are not altered themselves in the process. They work by lowering the amount of energy needed to make a neurotransmitter from its component molecules. Once the neurotransmitter is fashioned, the enzyme emerges unchanged and ready to participate in the formation of other neurotransmitters.)

The three other major classes of neurotransmitters are present at far fewer synapses. The *monoamines* are divided into two major classes: the catecholamines, which include epinephrine, norepinephrine, and dopamine and share a common chemical structure, and the indoleamines, which are synthesized from the amino acid tryptophan and include serotonin and melatonin. All of these are present in small groups of neurons that are primarily located in the brain stem. From these centers, elongated and highly branched axons ascend and descend to diffuse and often widely dispersed terminal points. *Acetylcholine,* the major neurotransmitter at the junction between nerves and muscle, also originates in subcortical structures above the brain stem. Finally, there is a chemically heterogeneous group of neurotransmitters that includes histamine, nitric oxide, and the neuropeptides—small molecules composed of short chains of amino acids. Among the neuropeptides are the endorphins (discussed in Chapter 14), and Substance P and Substance K (discussed in Chapter 15).

Dopamine is concentrated in a dark-appearing area in the brain stem that bears the exotic-sounding name substantia nigra; in Parkinson's disease nerve cells in this area are lost, resulting in a deficiency of this neurotransmitter. Norepinephrine-containing nerve cell bodies are prominent in another brainstem nucleus with a name that sounds gemlike, the locus ceruleus. Like dopamine, cells from the locus ceruleus, termed noradrenergic, are thought to be involved in depression, euphoria, and anxiety. Serotonin is formed in cells found in a nucleus whose name sounds like a commuter train: the midline (or dorsal) raphe. Fibers from these cells, as well as those from the noradrenergic cells of the locus ceruleus, are widely distributed throughout the brain. Serotonin plays a major role in the sleep-wakefulness cycle as well as in the biology of mood, emotion, and other functions mediated by the limbic system.

Thanks to the diffuse distribution of dopamine, norepinephrine, and serotonin, their action may affect large numbers of cells in different parts of the brain. Moreover, because these neurotransmitter systems originate in one small area in the brain stem, an injury to that area can have devastating consequences, affecting movement, thought, and emotion. The process is similar to what happens when a milk distributor goes on strike: The absence of trucks leaving the point of distribution produces panic throughout the city among families with children.

Acetylcholine, which also originates in subcortical structures above the brain stem, is the only major neurotransmitter that is not derived directly from an amino acid. It is also the only neurotransmitter that was discovered in a dream. In 1921 its discoverer, Otto Loewi, was searching for an explanation of why a frog's heart slows up when the vagus nerve in the neck is electrically stimulated. While he was asleep, Loewi dreamed of taking the fluid surrounding a frog's heart and applying it to the surface of another frog's heart, whose vagus, unlike that of the first frog, had not been electrically stimulated. When he carried out the experiment, Loewi discovered that the second frog's heart slowed too. From this he correctly deduced that the vagus nerve slows the heart through the release of a chemical. That chemical turned out to be acetylcholine. It is employed at many synapses throughout the brain, but most of the neurons that synthesize it are highly concentrated in an area in the lower part of the basal ganglia named after its discoverer, the nucleus basalis of Meynert. Acetylcholine plays an important role in memory; this area, which projects to much of the cerebral cortex, degenerates in Alzheimer's disease, producing a cortex that contains much less acetylcholine than normal.

The neuropeptides also influence information transfer within the brain. Some are primarily involved with inhibition; others with excitation. Peptides function principally as modulators, somewhat like stops on an organ. Rather than producing stimulation or in-

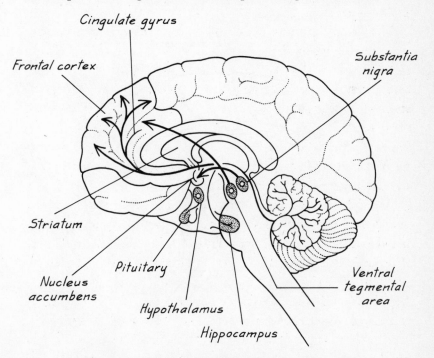

DOPAMINE PROJECTION SYSTEM IN THE BRAIN

hibition on their own, they facilitate such action in other neuro-transmitters. A neuron may employ one or more modulatory neuropeptides along with any of several neurotransmitters.

The chemical diversity provided by all these substances allows brain cells to exhibit a great flexibility and subtlety of response. In fact, one of the most exciting discoveries in brain research in recent years has been this discovery that neurons may utilize several neurotransmitters rather than a single one. This suggests that each neuron is capable of a greater variation in response than was considered possible only a few years ago. It is now believed that this wide repertoire of responses at the molecular level forms the basis for the rich behavioral complexity of our lives.

Consider our sense organs. In contrast to thermostats, thermometers, and light meters, the organs responsible for seeing and hearing and tasting and touching were not designed with fixed calibrations to give absolute readings of those aspects of our world

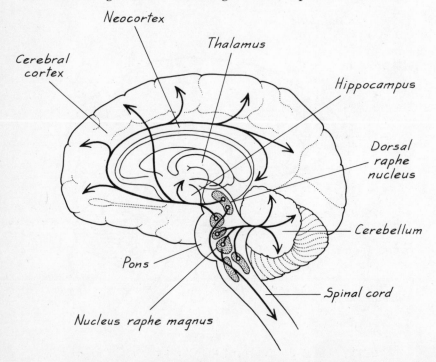

SEROTONIN PROJECTION SYSTEM IN THE BRAIN

that they monitor. Instead, these organs are designed to detect changes, often of exquisite subtlety. The retina within the human eye contains more than twenty different neuroactive agents. Present evidence suggests that the majority of these agents act by modulating retinal cells' responses to a neurotransmitter that excites or inhibits. Instead of responding straightforwardly to that transmitter's action, a cell's response resembles what happens when a photographer inserts a series of filters over his lenses. The pictures that result from this process all show the same scene, but they are different in important ways; subtle variations in shading and color distinguish one from another.

Within the brain, neurotransmitters and their receptors function like the lenses and filters of the photographer. Those neurotransmitters that stimulate or inhibit a nerve cell's firing correspond to the lenses. Depending upon the power of the lens selected and the direction in which the camera is pointed, the photographer has

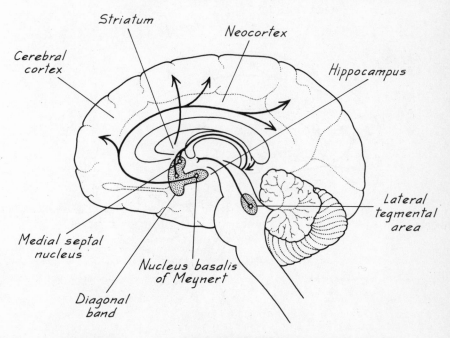

ACETYLCHOLINE (CHOLINERGIC) PROJECTION SYSTEM IN THE BRAIN

the potential to take an infinite variety of pictures with it. Now add to the lens a filter, which corresponds within the brain to a neurotransmitter that functions as a modulator. Just as a filter modulates the tone and quality of the resulting photograph, a neuromodulatory transmitter modifies a neuron's basic response in some way.

Moreover, neuromodulation is not confined to the cellular level—it occurs at every level of brain functioning. Think back to the last time you took an antihistamine or a sedative, or perhaps had a bit too much to drink. The chemical involved modulated your brain's responsiveness, and as a result you became tired, irritable, inattentive, or otherwise significantly different, "out of character." This neuromodulation at the behavioral level was mirrored at the molecular level: The drug or alcohol altered one or more of the neurotransmitters or their receptors. In this sense the brain is a self-reflecting organ; the most obvious and observable

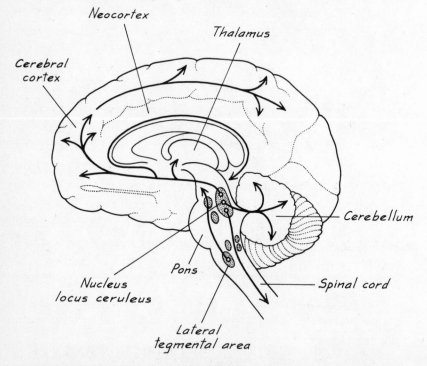

NOREPINEPHRINE (NORADRENERGIC) PROJECTION SYSTEM IN THE BRAIN

order of organization—human behavior—can be discerned and described at the level of its neurotransmitters and their receptors. This is because both human behavior and neurotransmitter activity involve the transfer of information from sender to receiver. It doesn't matter whether that information is a landscape that a photographer captures, the notes of a Brahms symphony, or the fragrance of a new perfume. *All* nerve impulses are the same, whatever informational messages they may be carrying; and brain activity at every level, from the anatomical to the molecular, can be thought of in terms of messengers and receptors.

At every level these messengers and receptors underpin the behavior by which we define ourselves. This holds true for everything from the twitch of an isolated muscle cell to the formulation of a unified field theory in physics. Behavior, defined in the dictionary sense as a "course of action," therefore occurs at multiple levels. As I am writing these words, for example, the aroma of freshly baked cookies is wafting upward from the kitchen to the gallery where I sit. A scientific explanation for what is going on would go something like this: The molecules responsible for the aroma stimulate receptors within my nose. Within the membranes of these nerve cells, which are specialized to detect a specific form of energy, chemical reactions take place. Only these specific cells and no others can carry out the chemical transformations necessary to detect the cookie aroma. If I am suffering from a head cold and have temporarily lost my capacity to smell, the nerve cells in my eyes or ears cannot take over for those in my nose; each sense is receptive only to specific stimuli.

But there is a puzzle here: If you were to compare nerve cells from different parts of the body—say, one that mediates sound and one devoted exclusively to touch—you would discover that these functionally divergent cells operate through the same mechanisms. These action mechanisms prevail throughout the brain, yet we perceive a world that is infinitely varied and rich. How can the brain do this?

The brain's marvelous performance is possible because of its organization. Indeed, it is best thought of as a configuration of receptors that are organized like nested Chinese boxes. All of them consist of the same materials, and all employ similar principles of construction. We speak here of self-reference; each level of organization reflects and is reflected by all the others. But starting with the largest and moving inward, each box is a smaller, more compact version of the ones before.

Think of what happens when you sit at a dressing table with mirrors in front of and on either side of you. Each time you turn or look to the right or left, you encounter an infinite series of images of yourself in descending scale, from a "true-to-life" size down to tiny images that seem to fade off into infinity.

Nowhere is such self-reference more evident than in the various levels of the human brain. At the *behavioral* level, we express ideas that originate in our cerebral cortex, through spoken and written language, gesture, and even the thoughtful employment of silence.

This behavioral activity is reflected on the *microscopic* level, where untold numbers of the brain's two to three hundred billion neurons are activated in intricate patterns that a neuroscientist once whimsically compared to an enchanted loom. Each cell communicates with one thousand to ten thousand other cells, making the total number of possible interactions exceed the number of particles in the known universe. The brain's self-reflection and self-reference continue still deeper, to the *molecular* level of organization.

Thus, when we are conversing with a friend, our neurotransmitters stimulate, inhibit, or otherwise exert control over our brain communication in ways that reflect the three levels simultaneously: the words, the accompanying neuronal firing patterns, and the biochemical, bioelectric, and magnetic field forces within the living brain.

At every moment, in time frames ranging from milliseconds to the eight or nine decades of the human life span, each of the three levels is related to all the others. Underlying this relatedness are receptors: specific behavioral, neuronal, and molecular structures that receive information from the outer world and transform it by means of symbols. The fact that our brain possesses the most elaborately evolved memory structure on earth makes it possible for us to perceive and articulate continually finer and subtler distinctions concerning events in our lives—distinctions that shape our attitudes and behavior over a lifetime. Symbolic processes occur at all levels of our functioning.

This is a notion that many people find takes some getting used to. While we are accustomed to using signs and symbols at the behavioral level—words, for example, clearly symbolize ideas, things, and events—the interactions of neurotransmitters with receptors seem far removed from the realm of symbols and signs. But neurotransmitters and their receptors also mirror the internal and external environment and thus serve as symbols and signs. Furthermore, symbols and signs—and transmitters and receptors of signs—are not limited to creatures like ourselves with highly evolved brains; they can be encountered all around us. "The entire universe . . . which we are all accustomed to refer to as the 'truth' . . . all this universe is perfused with signs if it is not composed exclusively of signs," wrote the eminent American philosopher C. S. Peirce.

It may seem simplistic to equate a disordered thought with a disordered molecule, but there are situations in which they seem literally to be equatable. When communication is disrupted at one level—say, when a crucial chemical in the brain undergoes an alteration—the end result of that event may be invisible to our eyes but will be only too "visible" many orders away. An infant born without the ability to metabolize an essential chemical in her food may, as a result of ensuing brain damage, spend her days staring into space, incapable of responding to those around her.

Perhaps it would be truer to say that a disordered thought is only a disordered molecule considered from a different perspective. Our vast new understanding of brain chemistry, that thoughts and emotions are the result of chemical processes going on within the brain, has given us the capability to deliberately alter that chemistry.

Drugs can now be used to manipulate neurotransmitters and receptors with subtlety and precision. But mind alteration has an ancient history, so rather than proceed directly to the laboratories of modern psychopharmacologists, it will be illuminating to start our journey several hundred years earlier, in the primeval jungles of the western Amazon and along the Pacific coastal areas of Colombia and Ecuador.

CHAPTER 3
MAGIC INTOXICANTS

A magic intoxicant known as ayahuasca ("vine of the soul") is employed in the jungles of South America and throughout the western part of the Amazon basin. This drink is believed to transport its subjects into wondrous realms where the soul, released from the sometimes tedious realities of everyday life, may encounter long-dead ancestors and other marvels.

The hallucinogen responsible for this effect is prepared by scraping bark from the freshly harvested stems of two plants, *B. caapi* and *B. inebrians.* The bark is boiled for several hours, and the bitter thick liquid that results is then swallowed in small quantities.

An initial period of dizziness, nausea, and sometimes vomiting is followed by lassitude and colored visual hallucinations: white progressing to a hazy, smoky blue, purple, or gray. The effects of ayahuasca are greatly enhanced when the leaves of two other plants, *B. rusbyana* and *B. psychotria,* are added to the mixture; the visual hallucinations turn to brightly colored reds and oranges.

The effects of ayahuasca on human perception and behavior have been known for centuries among the natives in South America. But its seemingly magical properties were explained only when specimens of the bark that had been stored since 1852 in the museum at the Royal Botanic Garden at Kew, in England, were subjected to biochemical analysis. In that year the botanist Robert Spruce had attended a feast given by Tukano Indians living close to Brazil's Amazonian frontier with Colombia. As part of the feast the Tukano drank caapi, a nauseous, bitter-tasting liquid that induced hallucinations. In *Notes of a Botanist on the Amazon and Andes,* Spruce wrote: "I learnt that Caapi was cultivated . . . a few hours journey down the river and I went there one day to get specimens of the plants, and (if possible) to purchase a sufficient quantity of the stems to be sent to England for analysis." His quest was successful, and he sent back the materials he had gathered. But neither Spruce nor any of his contemporaries lived long enough to learn the chemical ingredients of the "vine of the soul." Not until 117 years later, in 1969, was a chemical analysis of Spruce's materials finally carried out.

The bark of the plant contains the alkaloid harmine. Subsequent investigation revealed that almost all hallucinogenic plants, with the notable exception of cannabis (hashish or marijuana), perhaps the most widely used mind-altering drug in the world, contain nitrogen and belong to the chemical class of alkaloids. Chemical analysis of the constituents of other "sacred plants" led to an astonishing finding: The plants' principal hallucinogens are closely related in their chemical structure to neurotransmitters found in the human brain. Put another way, the plants' psychoactive agents bear a remarkable resemblance to agents involved in the biochemistry of mental function.

Peyote, for example, is a cactus with mind-altering properties that the Aztecs and Indians of Mexico have used for at least three thousand years. Its active principle is mescaline, a compound that is structurally similar to that of the brain neurotransmitter norepinephrine. Both mescaline and norepinephrine are related to

phenylethylamine, one of the four main categories of hallucinogens. (The other three categories are tryptamine, the ingredient in
ayahuasca; lysergic acid, or LSD; and the cannabinols, the ingredients in marijuana.) Indeed, all the major plant hallucinogens
contain substances similar in chemical composition to psychically
active chemicals within the brain.

In their book *Plants of the Gods,* Richard Evans Schultes, Jef-
frey Professor of Biology and director of the Botanical Museum
at Harvard University, and Albert Hofmann, the discoverer of
LSD, suggest that the modification of brain receptors on both the
molecular-chemical and the experiential level may account for the
powerful psychic effect of hallucinogens:

> If we adhere to the concept of reality as the product of the
> interaction between sender and receiver, the perception of a
> different reality under the influence of hallucinogens may be
> explained by the fact that the brain which is the site of con-
> sciousness undergoes dramatic biochemical changes.

Extending their vision from molecules interacting at the submi-
croscopic level to the most intriguing macromolecules in the
world, ourselves, Schultes and Hofmann continue: "The receiver
is thus set for wavelengths other than those associated with normal
everyday reality." They suggest that by bringing about "biochemical modifications of the brain field," it is possible to "produce
changes in the awareness of reality . . . this constitutes the real significance of hallucinogens."

Not all attempts to alter reality by means of biochemical manipulation involve glorious visions and ecstatic states, however. A
woodcut attributed to Hans Holbein depicts a medieval witch
standing in her garden of hexing herbs. Before her is an arbore-
tum of substances being cultivated either for their intoxicating
properties or for their bizarre shapes; the root of the mandrake,
for instance, is curved, gnarled, and twisted into a shape resembling the human body.

The most famous constituent of the witch's garden is the deadly nightshade, or atropa belladonna. The double name reflects two of its uses in the Middle Ages. In Greek mythology Atropos was the oldest of the three Fates; an inflexible and detached figure, it was she who cut the thread of life at the critical moment. Over the centuries both amateur and professional poisoners willingly took on the duties of Atropos and employed the deadly night-shade to dispatch the unwanted and the inconvenient.

Belladonna, the plant's species name, roughly means "beautiful lady," a reference to the use of the extract of this plant to dilate the pupils of the eyes. Egyptian and Roman women are credited with this discovery. Its activity was confirmed by scientists in the 1950s, when studies revealed that most people, when shown paired photographs of women who differ only in the amount of pupillary dilation, select as more appealing the woman with the more widely dilated pupils.

But during the Middle Ages atropa belladonna was used for purposes other than poisoning rivals and beautifying ladies. It was a primary ingredient in a mixture capable of inducing the sensation of flying, or levitation. In preparation for the Sabat, or Black Mass, an ointment would be prepared; a typical recipe might consist of the fat of a stillborn child, juice of water parsnip (not the kind we eat today but hemlock, the poison forced upon Socrates), aconite (monkshood, the poisonous plant that killed Romeo), belladonna in combination with two other hexing plants of the family Solanaceae: the "herbs of consolation" mandrake and henbane (the name of the latter indicates that the substance is poisonous to hens), and finally soot, included in order to render the mixture black and therefore suitable for camouflaging the body at night.

This psychoactive paste was made into a salve and rubbed on the body, especially against the back of the thighs and most lib-erally between the legs and on a stick that could be straddled or inserted into the vagina. The now-familiar witch's broomstick was graphically described in a 1324 investigation into witchcraft: "In rifleing the closet of the ladie, they found a Pipe of oyntment

wherewith she greased a staffe, upon which she ambled and gal-
loped through thick and thin, when and in what manner she
listed."

It may seem a long journey from a fourteenth- or fifteenth-
century orgy to a 1990s study of the brain, but many reported
effects of the witch's ointment are explicable today in terms of
neurotransmitters and receptors. All the hexing herbs contain in
different concentrations the same active principles: the alkaloids
hyoscyamine, atropine, and scopolamine. All are powerful blockers, or antagonists, of a receptor within the brain for the neurotransmitter acetylcholine. These "anticholinergic" agents occupy
the acetylcholine receptors but do not activate them; they thereby
block or "antagonize" the normal activity of the parasympathetic
nervous system.

A gradual increase in the dose causes additional peripheral
blockage, which induces dryness of the mouth, a decrease in per-
spiration and a resulting elevation in temperature that in the case
of severe atropine poisoning may reach 107 degrees, and a racing
of the heart, sometimes exceeding a fifty-beat-per-minute increase.
At low to moderate doses these chemicals cause pupillary dilation
that makes it impossible to focus the eyes on nearby objects. At
higher doses they exert their effects on the brain and the central
nervous system and result in psychosis and other forms of agitation and confusion—one reason why, when given therapeutically,
their dosage must be carefully controlled.

Poisoning with alkaloids such as atropine, hyoscyamine, and
scopolamine induces hallucinations of a peculiar and highly specific quality. Unlike hallucinogens like LSD or mescaline, the anticholinergic hallucinogens produce intoxication, accompanied by
a clouding of consciousness and loss of memory for the period of
intoxication. "The user remembers nothing experienced during
the intoxication, losing all sense of reality and falling into deep
sleep like alcoholic delirium," write Schultes and Hofmann.

This comparison with alcohol's ability to induce amnesia is apt.
Both acute and chronic alcohol intoxication induce damage in

parts of the brain concerned with memory, especially the ability to set down new memories. After a night of heavy drinking, it is not unusual for the imbiber to wake up not only with a crushing hangover but also with a total absence of memory of all the inappropriate things done the night before. If the drinking becomes habitual, the capacity to set down new memories may be permanently lost. A similar situation occurs with the acetylcholine receptor blockers contained in the hexing herbs, suggesting that the same brain areas are involved.

The effects of blocking the acetylcholine receptors yield a plausible explanation for what occurred during the witches' Sabat. The feeling of levitation or flying came from the combination of an irregular heartbeat, drowsiness, the psychedelic effects of the alkaloids (especially scopolamine), and the power of suggestion, or what we would now label mass hysteria. Combine the pounding and racing of the heart with the mildly aphrodisiac properties of belladonna, and it is little wonder that participants in nightlong sexual orgies felt able "to be carried in the aire, to feasting, singing, dansing, kissing, culling and other acts of venerie with such youthes as they loue and desire most."

The next day, all memory of the night's activities would be gone. Those in the Middle Ages who so vehemently denied accusations of witchcraft and participation in the Sabat may have been speaking truthfully when they said they did not remember what occurred. Torture and other means of persuasion were no doubt applied in the hope of inducing the accused to admit to the actions attributed to them by eyewitnesses. But thanks to the amnesiac effect of the scopolamine, the accused would be truly unable to recall the orgy and therefore were likely to deny participation.

Similar amnesias occur as a side effect of current drugs that act as antagonists to the acetylcholine (cholinergic) receptors. When scopolamine is given to volunteer subjects, it reduces their verbal free recall and interferes with their sustained memory. Subjects under its influence have a harder time learning a list of words, mistakenly come up with many words that are not on the list, and correctly recall fewer list words than subjects who are not given the drug.

Increasing doses of scopolamine and other acetylcholine receptor blockers (usually in the form of over-the-counter cold remedies) may produce behavior that in earlier times would have been ascribed to possession: widely dilated eyes, weakness and clumsiness, mental confusion, speech disturbances, visual and auditory distortions, and hallucinations. At still higher doses come thought blocking, disorganization, feelings of strangeness and alienation from one's own body, loss of contact with reality, and the diminished capacity to distinguish between objective events and disturbances in one's own inner processing.

Curiously, the toxic effects of cholinergic antagonists have been deliberately induced to treat certain mental illnesses, especially mania and schizophrenia. In the 1950s some psychiatrists employed atropine coma therapy, a form of acetylcholine receptor blockade. Increasing doses of atropine or scopolamine were injected until a dose was eventually reached that induced coma. Six to eight hours later the patient awakened—according to the method's proponents, greatly improved. The symptoms leading up to the therapeutic coma are exactly the same as those observed in the medieval Sabat: flushing, pupillary dilation, dryness and burning of the eyes and mouth, talkativeness, restlessness, incoordination, disorientation, and visual hallucinations. This bizarre and outmoded therapy is an example of the arbitrariness of separating a medicine from a poison, an intoxication from a therapeutic treatment. "A poison is often nothing more than a medicine wrongly prescribed," as a physician once put it.

All the chemicals discussed in this book possess the potential for both benefit and harm. It matters a great deal how and why and under what circumstances a mind-altering substance is employed. The plant *Hyoscyamus niger* or henbane, for example, has been used at various times as a human poison and as a vehicle for achieving prophetic powers. Homer wrote of magic drinks containing the drug; some historians even suggest that the priestess at the Temple of Delphi prophesied while under the influence of smoke from henbane seeds. Both Hamlet's father and the Roman emperor Claudius were killed by surreptitiously supplied henbane. During the Middle Ages henbane served as a means of conjuring up demons. In epochs marked by executions, the drug's pain-killing properties were used to relieve some of the suffering.

But the most frequent use of henbane was as an adjunct to sexual orgies and satanic rituals. Dr. Robert S. De Ropp, in *Drugs and the Mind,* captures the spirit of this aspect of the substance:

> At the Bacchanalia, when the wild-eyed Bacchantes with their flowing locks flung themselves naked into the arms of the eager men, one can be reasonably certain that the wine which produced such sexual frenzy was not a plain fermented grape juice. Intoxication of this kind was almost certainly a result of doctoring the wine with leaves or berries of belladonna or henbane. The orgiastic rites were never totally suppressed by the Church and persisted in secret forms through the Middle Ages. Being under the shadow of the Church's displeasure, they were inevitably associated with the Devil, and those who took part in them were considered to be either witches or wizards.

Unless we are mindful that a poison is often a medicine wrongly prescribed, many of these uses would seem incompatible. How can a substance that serves as a poison also serve as an intoxicant? The poet Pliny suggests the answer: "For this is certainly known, if one takes it in drink more than four leaves, it will put him beside himself." Pliny might have added that if substantially more than four leaves are taken, this insanity will proceed to death.

In short, dosage is the key to the plant's effects. In mild doses its effects may be harmless, even beneficial, as when it is used as a cold remedy. In larger doses a pleasant "high" may give way to madness; and at a still higher dose, prostration and death may result. Mind- and brain-altering substances involve a similar dose dependency.

Several possible mechanisms may explain this. Some brain areas may take up a drug more avidly than others. Receptors in these areas would take up all the drug when the amount consumed is low. As the quantity increases, these receptors are filled, and the drug may then be taken up by other brain areas that have slightly less specific receptors. Think of it as a sale where, say, pre-Castro Havana cigars will go on the market at a certain hour in a specific store. Those who arrive first and buy up the initial offering are likely to be individuals with the greatest avidity—receptivity—for fine cigars. But if enough of them are offered for sale and the price becomes low enough, eventually people who ordinarily aren't "into" cigars will buy them. People might even try the cigars who haven't smoked before. It is they who are likeliest to get sick.

The use of plant-derived mind-altering drugs continued for several hundred years, and depending on the preconceptions and prejudices of the time, the users were considered to be prophets, seers, madmen, mystics, or followers of the Devil. The explanations for users' behavior emphasized at different times and to varying degrees darkness, the powers of "evil," the demonic, and the depraved. Such interpretations continued until the waning years of the nineteenth century, when a pioneering researcher's interest in an ancient plant forever changed our ideas about the human mind.

CHAPTER 4
DR. HOFMANN AND THE MAGIC CIRCLE

In 1896, Louis Lewin, a German chemist and expert on poisons, journeyed to the United States to collect specimens of the hallucinogenic peyote cactus for chemical analysis.

Lewin occupies a pivotal position among researchers of the chemistry of the human brain. Pictures of him show him looking remarkably like one of the cough-drop Smith Brothers: a bearded, scholarly looking man in a dark suit and high starched collar. His facial expression suggests intelligence, curiosity, and an analytic disposition, coupled with a Gene Wilder kind of zaniness. Lewin has been called, with good reason, perhaps the most influential figure in the history of psychopharmacology: the science that aims at influencing the mind through chemistry.

His reputation depends for the most part on *Phantastica: Narcotic and Stimulating Drugs,* published in 1924. Since its English publication in 1931, this book has become a classic. It made prescient statements about the power of drugs to influence the brain: "The strongest inducement to a frequent or daily use of the substances in question is to be found in the properties they possess; in their capacity to excite the functions of the brain-centres which transmit agreeable sensations and to maintain for some time the consciousness of experienced emotions."

Lewin's greatest accomplishment may have been his classification of narcotic and stimulating drugs. Until his book appeared, most people's, even most scientists' understanding of the interactions of drugs, mind, and behavior consisted of a mélange of misinformation, prejudice, value judgments, and superstition. Lewin dispelled all of this through his recognition of the reasons people take drugs. He spoke of "classifying the agents capable of effecting a modification of the cerebral function, and used to obtain at will agreeable sensations of excitement or peace of mind."

It seems clear that, given a choice, most people most of the time, for different reasons and on different occasions, desire either to be in a stimulated, heightened state of mind or to experience a sense of tranquillity, peace, and contentment. Personality—in what Lewin refers to as "the toxic equation," or individual susceptibility to a chemical—determines the appeal of a particular drug, whether it is an excitant like cocaine or a euphoriant like heroin.

Any classification that dates back more than sixty years must leave a lot to be desired: Although tranquilizers and mood elevators are now common, for example, they were unknown in 1924. Nevertheless, Lewin's concept of the social, behavioral, and experiential effects of drugs remains useful and revealing.

Lewin divided mind- and brain-altering drugs into five categories, which can be depicted as separate spokes emanating from a central axis. He made no claim that all drugs can be neatly assigned to only one of these five categories; many drugs have multiple effects. Even a single person may respond differently to the same drug at different times depending on the occasion, his or her expectations, the previous use of mind-altering substances, and other factors.

Lewin's first category consists of the *euphoriants*. These substances diminish or even suspend unpleasant or disturbing emotions or perceptions, while they usually produce little effect on consciousness. They induce a state of mental and physical comfort. Thanks to euphoriants, the difficulties and frustrations of the present can be replaced by substitute worlds of the mind's own creation, worlds in which problems disappear, anxieties are quieted, and desires are sated. Included in this category are opium and its derivatives heroin and codeine. (Lewin also placed cocaine in this category, although given what we have learned about this drug over the past seventy years, most contemporary experts would assign it to the excitants.)

The *phantasticants,* also termed drugs of illusion, induce cerebral excitation in the form of illusions, visions, and other hallucinations, Lewin believed. He referred specifically to mescal buttons (*Anhalonium lewinii,* named after himself), dried pieces of the gray-brown spineless crown of the peyote cactus that con-

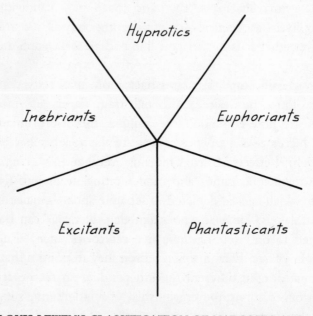

**LOUIS LEWIN'S CLASSIFICATION OF NARCOTIC AND
STIMULATING DRUGS**

tain the psychoactive alkaloid mescaline. In 1897, Lewin and another chemist now largely forgotten, Arthur Heffter, isolated mescaline from the parent cactus. It would be another twenty-three years before a Viennese chemist, Ernst Spath, succeeded in analyzing the chemical structure of mescaline and synthesizing it in the laboratory.

The *inebriants,* Lewin's third group, produce intoxication and are the most commonly abused of the mind-altering drugs. Typically, behavioral excitement is followed by disturbances in perception, thinking, and emotion. Depression then follows. The inebriants can be produced by chemical synthesis and include chloroform, ether, and—the most extensively employed mind-altering substance in the world—alcohol.

Hypnotics, the fourth group, are sleep-producing agents. Depending on dosage, sleeping pills can also lead to amnesia and confusion. Like drugs in the other four groups, hypnotics induce in the user a retreat from the world—in this case, into a world of sleep and dreams.

The *excitants,* the fifth and final group, are the mental stimulants. Ranking not far behind the inebriants in popular appeal, these substances produce a more or less apparent excitation of the brain without altering consciousness. With brain excitation come behavioral arousal and stimulation. That first morning cup of coffee contains the naturally occurring excitant caffeine. Synthetically produced excitants include methamphetamine ("speed"), and methylphenidate, sold under the trade name Ritalin.

Even today, Lewin's classifications remain useful enough for us to refer to them as we explore brain and mind alteration through chemistry. His knowledge of this then-esoteric subject was encyclopedic. He knew more about narcotic and stimulating drugs than any person alive. If this had been the full extent of his genius, it would have been enough to earn him a preeminent place in the chronicle of twentieth-century men and women who employed chemical research to reveal secrets about the brain and mind. But Lewin's abilities extended well beyond those of a mere classifier.

Like all creative people, he employed his intelligence and imagination to discover underlying principles; he made certain leaps of faith, with a willingness to speculate and extend his vision beyond simple observation and classification. Two of his insights stand out.

First, Lewin recognized that all mind-altering substances act by affecting physical functions. "[H]ave visions and hallucinations a material cause?" he asked in *Phantastica.* "Yes, in my opinion. The nature of the cause may not always be the same, but it is always an excitation localized in the interior of the body." This was a radical assertion. Even as late as 1924, such behaviors were likely to be seen as dark, malevolent, and often supernatural. The Salem witch trials of 1692, the worst outbreak of witch persecution in American history, stand out as an extreme interpretation of bizarre and inexplicable behavior, and while it is true that few people still believed in witches more than two centuries later, exorcisms were still carried out with some regularity.

The second of Lewin's insights was that hallucinations are caused by chemicals working not just anywhere in the body but specifically within the brain. "If any light is ever to be shed on the almost complete darkness which envelops these cerebral processes, then such light will only originate from chemistry." This view of the chemical nature of brain action would prove prophetic.

✛ ✛ ✛

In the spring of 1943, Albert Hofmann, a chemist working at the Sandoz pharmaceutical research laboratories in Basel, Switzerland, began an intensive investigation of the amides of lysergic acid (LSD, etc.), the chemical backbone of the plant fungus ergot.

A highly poisonous mold under certain climatic conditions—cold winters and wet, warm summers—ergot can grow rapidly and infect rye. When eaten in toxic doses, ergot-infected rye can produce gangrene, a painful constriction of the blood vessels that interferes with their ability to deliver oxygen to body tissues. If severe and sufficiently prolonged, gangrene can lead to the loss of the nose, the earlobes, fingers, toes, and in extreme cases entire limbs.

At lesser doses, ergot's vessel-constricting effects are not only much less powerful, they can be put to good use. Ever since the sixteenth century, midwives and obstetricians have employed the fungus to induce labor and limit blood loss. A newly developed chemical procedure allowed Hofmann and other scientists at Sandoz to combine lysergic acid derived from ergot with other chemicals; their efforts at synthesizing variations of natural ergot yielded drugs that are still in use today as mainstays in obstetrical practice. One such drug, ergonovine, is used to induce labor; another, methergine, arrests hemorrhage and is employed to cut down on blood loss during delivery.

But ergot's effects are not limited to the constriction of blood vessels. Outbreaks of panic, bizarre behavior, hallucinations, delusions, and epilepticlike convulsions that are secondary results of ergot poisoning can be traced back to at least the fourteenth century. Some historians suggest that the witch accusations at Salem may have resulted from ergot-containing rye crops and their ingestion by the poorer members of society, who could not afford the more expensive wheat-based breads.

By 1943, the chemists at Sandoz had synthesized a long series of compounds containing LSD. One of them was LSD-25, so called because it was the twenty-fifth compound of the series. Hofmann, intrigued by the possibility that it might serve as a brain stimulant, or analeptic, was convinced there should be "more profound studies with the compound." He prepared a fresh quantity of LSD-25 during the second week of April. On Friday of that week he was "seized by a peculiar restlessness associated with a sensation of mild dizziness." By midafternoon the feelings had increased to the point that he was forced to stop work and go home. Here in Hofmann's own words, in a letter to his supervisor dated April 22, 1943, is what happened next:*

> On arriving home I lay down and sank into a kind of drunkenness which was not unpleasant and which was characterized by extreme activity of imagination. As I lay in a dazed condition with my eyes closed (I experienced daylight as disagreeably bright), there surged upon me an uninterrupted stream of fantastic images of extraordinary plasticity and vividness and accompanied by an intense kaleidoscope-like play of colors.

Many people's response to such an experience—certainly mine—would have been a good deal of relief that things hadn't been even worse. Depending on one's personality and predispositions, one could come up with various interpretations of what had happened: the onset of insanity, a spiritual visitation, chemical poisoning. Quite sensibly Hofmann opted for the last:

> The nature and course of this extraordinary disturbance raised my suspicions that some exogenic intoxication may have been involved and that the lysergic acid diethylamide with which I had been working that afternoon could have been responsible.

Nonetheless, two considerations seemed to fly in the face of Hofmann's suspicion of chemical origins. For one, the nature of his experiences did not correspond to those associated with ergot poisoning. No convulsions, maniclike excitement, or paranoid fears or delusions occurred. Moreover, the experience taken as a whole was not unpleasant; Hofmann had remained calm and alert throughout. The mental effects of ergot poisoning, in contrast, are unremittingly unpleasant: fright, outright panic, and an inability to maintain a sense of distance in the face of such strange expe-

---

*Hofmann's accounts of these experiences are taken from his paper, "The Discovery of LSD and Subsequent Investigations on Naturally Occurring Hallucinogens," published in *Discoveries in Biological Psychiatry,* edited by Frank Ayd and Barry B. Blackwell (Baltimore, Md.: Ayd Medical Communications, 1984).

riences. These are the most distressing aspects of ergot poisoning, and none of them had occurred in Hofmann.

The second consideration concerned dosage. Hofmann had been working with only small amounts of the drug, and he could not account for how the substance could have found its way into his body in sufficient quantity to produce such stark psychological effects. (It seems likely that it was absorbed through the skin.) Ever the scientist, Hofmann decided "in order to get to the root of the matter . . . to conduct some experiments on myself with the substance in question." He started by taking the lowest dose of LSD he could conceive as sufficient to exert any effect: 0.25 milligram. At that time he had no way of knowing that he had grossly underestimated the drug's activity at even extremely low doses; 0.25 milligram, it turns out, is five times the average effective dose of LSD.

Fifty minutes after swallowing the tasteless dose, Hofmann's symptoms of three days earlier returned: dizziness, unrest, difficulty in concentration, and visual disturbances. In addition, this time he experienced a nearly irresistible impulse to laugh. He found that he couldn't continue taking notes because of these disturbances. After several fruitless efforts to write, he asked an assistant to accompany him home. We can assume that his coordination was not too poor since he and his assistant cycled home without incident, despite the fact that in Hofmann's case "my field of vision swayed before me and objects appeared distorted as in curved mirrors."

But by the time he arrived home, Hofmann knew that this second encounter with LSD was going to be much more powerful than the first:

> As far as I remember, the following were the most outstanding symptoms: vertigo, visual disturbances; the faces of those around me appeared as grotesque colored masks; marked motoric unrest, alternating with paralysis; an intermittent heavy feeling in the head, limbs, and entire body, as if they were filled with lead; dry constricted sensation in the throat; feeling of

choking; clear recognition of my condition, in which state I sometimes observed in the manner of an independent neutral observer that I shouted half insanely or babbled incoherent words. Occasionally I felt as if I were out of my body.

Reading this account almost half a century later, it is clear that Hofmann experienced a number of frightening, indeed terrifying reactions, much like accounts I have heard from people suffering from "bad trips." Typically such victims—this seems an appropriate choice of word—are completely out of control and are convinced of the reality of their hallucinations. If they are not restrained, they may carry out the most irrational and self-destructive acts in order to bring these highly unpleasant experiences to a conclusion.

Even after the immediate effects of the drug have worn off, certain aspects of the experience may recur in frightening and inexplicable flashbacks. Indeed, recurring states of perplexity and anxiety may become permanent aspects of the personality, as they did in a patient I treated for over a year. Curiosity about what happened, morbid brooding about quasi-mystical explanations of what can be recalled, nightmares, and phobic fears in regard to flashbacks—any or all of these consequences can seriously cripple the personality.

The fact that none of these things happened to Albert Hofmann again tells us something important about the effects of drugs on the mind: Context matters. Hofmann was not a thrill seeker but a scientist with great curiosity about what he had just experienced.

Retaining his perspective as "an independent neutral observer," Hofmann continued his description:

Six hours after ingestion of the LSD, my condition had already improved considerably. Only the visual disturbances were still pronounced. Everything seemed to sway and the proportions were still distorted like the reflection in the surface of moving water. Moreover, all objects appeared in unpleasant, constantly changing colors, the predominant shades being sickly green and blue. When I closed my eyes, an unending series of colorful, very realistic, and fantastic images surged in upon me. A remarkable feature was the manner in which all acoustic perceptions (*i.e.,* the noise of a passing car) were transformed into optical effects, every sound invoking a corresponding colored hallucination constantly changing in shape and color like pictures in a kaleidoscope. At about one o'clock I fell asleep and awoke the next morning feeling perfectly well.

The conclusion of Hofmann's ordeal marked a milestone in our understanding of the human mind and brain. Not that such experiences, journeys into the hinterland of the human psyche, hadn't been reported before. Similar descriptions existed in shamanistic-based writing over the centuries. Further, some of the stranger symptoms that Hofmann endured accompany schizophrenia. Schizophrenics frequently speak of feeling out of their body and experience sounds and other perceptions as somehow distinct from the ordinary. What made Hofmann's perceptions unique was not the content of his hallucinations but what the whole experience implied about the nature of the mind. In his own words:

> It lent support to the hypotheses that certain mental illnesses that were supposed until then to be of a purely psychic nature had a biochemical cause because it now seemed feasible that undetectable traces of a psychoactive substance produced by the body itself might produce psychic disturbances.

This was a profound insight, and Hofmann's experience had implications far beyond mental illness. It was now clear that a chemical—furthermore, a chemical synthesized in a laboratory and taken in only the tiniest amounts—could exert an overwhelming effect on psychic functioning. (Indeed, in terms of dosage, LSD's potency is unique.)

On the basis of his experience with LSD, Hofmann became interested in three major plants that the Aztecs and neighboring tribes used in their religious ceremonies. He consulted writings that dated from just after the conquest of Mexico by Cortés, and he discovered references to several plants, considered holy by the native population, that produced psychic effects. The first of these plants to arouse his scientific interest was peyote, the plant that had so interested Louis Lewin and from which he and Heffter had isolated mescaline. Mescaline made it possible for Hofmann to use a pure chemical compound to investigate hallucinogenic effects from a scientific, pharmacological, and clinical perspective.

Mescaline and LSD produce remarkably similar effects. While LSD is by far the more active and potent hallucinogen (somewhere between five and ten *thousand* times more active than mescaline), the nature and quality of the hallucinations induced by these two chemicals are remarkably similar. We know this from descriptions of peyote use in religious ceremonies in Mexico, Guatemala, Peru, and El Salvador over many centuries. Although no one is certain precisely how long these revered plants have been so employed, archaeological discoveries of peyote specimens in caves in Texas point to ceremonial uses more than three thousand years ago.

In late twentieth-century culture religious mysticism is largely absent as a motivating or explanatory force. It is easy for us to underestimate the reverence in which sacred plants were once held by indigenous cultures that discovered and employed them. These peoples literally worshiped them and used them only in the most sacred and solemn of their ceremonies. Little is known of these ceremonies today, thanks largely to the horror such practices aroused in the European missionaries, who launched stern and largely effective efforts to eliminate "idolatry."

But as often happens when religious practices are officially forbidden, their surreptitious use continued in underground settings. Apparently, the use of these plants was so well concealed that from the middle of the seventeenth century until the late 1930s, no botanist or anthropologist seems even to have suspected the religious use of teonanacatl, the second of the sacred plants, a mushroom whose Aztec name means "divine flesh." In 1938 a team of American investigators learned that the mushrooms were still being employed as magical potions in certain remote districts of southern Mexico.

But progress in learning about them was slow and cumbersome, thanks to the secrecy that surrounded these practices. Finally in 1955 came a spectacular breakthrough: An American investment banker and amateur investigator of psychoactive plants, R. Gordon Wasson, took part in a secret nocturnal ceremony in the Huautla de Jiménez, a village in the province of Oaxaca. On that night he became the first outsider known to ingest the holy mushrooms. Wasson described his experience in language unusual for a banker:

> It permits you to travel backwards and forward in time, to enter other planes of existence, even (as the Indians say) to know God. What is happening to you seems freighted with significance, beside which the humdrum events of every day are trivial. All these things you see with an immediacy of vision that leads you to say to yourself, "Now I am seeing for the first time, seeing direct without the intervention of mortal eyes."

A year later, Wasson returned to the area, this time in the company of mushroom expert Roger Heim, director of the cryptogamy laboratory of the National Museum of Natural History in Paris. Heim's aim was twofold. First, he wished to classify the sacred mushrooms botanically; he was able to place them in the family of Strophariaceae. He was also able to cultivate one of the mushrooms artificially in the laboratory—the genus Psilocybe mexicana Heim, named after him. But his second aim eluded the best efforts of both Heim and his fellows in Paris: to isolate the active principle within the mushroom that was responsible for its hallucinogenic powers.

Learning of Hofmann's discovery of LSD, Heim forwarded samples of the mushrooms to him at the Sandoz laboratories in Basel. This sequence of events was unique in the history of research on mind-altering substances. Typically, when a plant is discovered to possess psychoactivity, researchers attempt to modify the natural chemical in it or to synthesize an artificial analogue. But in this case, the progression was from an artificially synthesized drug to a natural substance. If Hofmann had not chemically synthesized LSD and published his findings, it's unlikely that Heim would ever have thought of sending him samples of the mushrooms, and we might still be ignorant of the psychoactive substances they contain.

First Hofmann tested the mushrooms' effectiveness on laboratory animals. Nothing very exciting occurred. Fearing that drying and culturing the mushrooms in Paris might have eliminated their potency, Hofmann decided to make another exploration into the inner realms of his own mind. After swallowing 2.4 grams—a medium dose—this is what happened:

> Thirty minutes after taking the mushrooms, the exterior world began to undergo a strange transformation. Everything assumed a Mexican character. As I was perfectly well aware that my knowledge of the Mexican origin of the mushroom would lead me to imagine only Mexican scenery, I tried deliberately to look on my environment as I knew it normally. But all voluntary efforts to look at things in their customary forms and colors proved ineffective. Whether my eyes were closed or open, I saw only Mexican motifs and colors. . . . At the peak of the intoxication, about one and a half hours after ingestion of the mushrooms, the rush of interior pictures, mostly abstract motifs rapidly changing in shape and color, reached such an alarming degree that I feared that I would be torn into this whirlpool of form and color and would dissolve. After about six hours the dream came to an end. Subjectively, I had no idea how long this condition had lasted. I felt my return to everyday reality to be a happy return from a strange, fantastic, but quite real world to an old and familiar home.

Contrast Hofmann's description with that of Maria Sabina, the contemporary shamaness of Huautla de Jiménez who uses the mushrooms in healing ceremonies:

The more you go inside the world of teonanacatl, the more things are seen. And you also see our past and our future, which are there together as a single thing, already achieved, already happened. . . . I saw stolen horses and buried cities, the existence of which was unknown, and they are going to be brought to light. Millions of things I saw and I knew. I knew and saw God: a vast clock that ticks, the spheres that go slowly around, and inside the stars the earth, the entire universe, the day and the night, the cry and the smile, the happiness and the pain. He who knows to the end the secret of teonanacatl can even see that infinite clockwork.

Both descriptions—the scientist's and the shaman's—share common features: the sense of being transported into other realms where knowledge is intuited rather than acquired in the usual linear, time-dependent pattern, and the experience of vivid colors and abstract patterns that alternate with visions of mythical and cosmic origin. The similarities in the experiences of such utterly different observers suggest that the active ingredients of the magic mushrooms exert their effects in an area of the brain that integrates emotions, colors, forms, and the sense of personal identity. The most likely candidate for this is the limbic system, especially those areas within the temporal lobe. Temporal lobe epileptics such as Fyodor Dostoevsky have described less intense but essentially similar experiences just before the onset of an epileptic seizure.

Hofmann successfully isolated the active principles of teonanacatl. Its main active component was psilocybin, and an alkaloid, psilocin, was present in trace amounts. Both are derived from the same compound, tryptamine. Tryptamine is found in the neurotransmitter serotonin, which plays a prominent role in the biology of mood, emotion, and other functions mediated by the limbic system. Identification of the active principles was soon followed by the synthetic production of psilocybin and psilocin in the laboratory. Chemical synthesis is preferable for several reasons; not only is it more economical, but the amounts of the active ingredients can be more precisely controlled. In the mushrooms themselves the main active principle tends to vary between 0.1 and 0.6 percent of the dry weight of the plant tissue. The user has to ingest far less of the synthesized substance to achieve a psychedelic effect: about 0.008 gram of synthetic psilocybin, as compared with two grams of the unpleasant-tasting mushroom.

In 1963, Hofmann attended a nocturnal mushroom healing ceremony in Mexico and supplied Maria Sabina with some pills containing synthetic psilocybin. She swallowed them instead of the mushrooms just before the ceremony. "It was a gala performance," wrote Hofmann in his memoirs. "At dawn when we left the hut, our interpreter told us that Maria Sabina had said there was no difference between the pills and the mushrooms. This was the final proof that the synthetic psilocybin was identical in every respect with the natural product."

The achievement of chemical synthesis, some might claim, took much of the mystery and fascination of the hallucinogen away. The active principle turned out to be not a god residing within the mushrooms, a view that dated back thousands of years, but two chemicals that could be manipulated at the whim of a chemist. To this extent, the activity of the mushrooms had indeed been explained. But as often happens in neuroscience, what was achieved was not so much an explanation as a shift of emphasis from one category to another. In this instance, myth yielded to chemistry; instead of speaking of gods and sacred substances, we speak of psilocybin and psilocin. Nonetheless, mystery remains. "It should be remembered that scientific investigation has merely shown that the magic properties of the mushrooms are the properties of two crystalline compounds. Their effect on the human mind is just as inexplicable and just as magical as that of the mushrooms themselves," wrote Hofmann and Schultes in *Plants of the Gods.*

✛   ✛   ✛

Hofmann's synthesis of psilocybin laid bare the secret of the second magic plant from Mexico. Still awaiting investigation was the third, ololiuqui. The first description of this plant is from the sixteenth century, when Francisco Hernandez, a Spanish physician with an interest in the flora and fauna of Mexico, prepared a report for King Philip II. In the Latin version that appeared in Rome almost a century later, in 1651, Hernandez described ololiuqui as "a snake plant . . . a tiny herb with thin green cordate leaves, slender green terete stems, and long white flowers. Formerly, when the priests wished to commune with their gods and to receive a message from them, they ate this plant to induce a delirium—a thousand visions and satanic hallucinations appeared to them."

With the help of a colleague, Hofmann obtained a sample of ololiuqui from a Zapotec Indian who lived near Oaxaca. Hofmann's chemical analysis of it produced unexpected results. The active principle of ololiuqui consisted of some familiar substances: lysergic acid derivatives. "Thus in this strange Mexican drug we met with old friends, since lysergic acid derivatives and ergot alkaloids had been favorite subjects of research in our laboratory since the time I had first synthesized LSD in the nineteen thirties," Hofmann recalls.

The active chemical in ololiuqui differs from LSD by virtue of a minor but critically important feature. In LSD, two ethyl groups $(CH_2 CH_3)$ replace two hydrogen atoms found in ololiuqui.

Otherwise, the two substances are chemically identical. Far from being insignificant, however, the substitution of two ethyl groups for the hydrogen atoms radically changes the psychic effect the drug produces. Here is Hofmann's description of what happened after he took two milligrams of lysergic acid amide, the main component of ololiuqui: "I experienced tiredness, apathy, a feeling of mental emptiness and of the unreality and complete meaninglessness of the outside world."

This psychic effect was in sharp contrast to the unending series of colorful, very realistic, and fantastic images induced by LSD, and from the "rush of interior pictures, mostly abstract motifs rapidly changing in shape and color" invoked by psilocybin.

An irony in all this both amazed Hofmann and delighted him: Lysergic acid amide was none other than LA-111, a laboratory code name for an experimental drug he had synthesized earlier during his investigations of LSD and related compounds. As unlikely as it seemed, his earlier laboratory manipulations had resulted in the synthesis of a compound that nature had hidden in an obscure plant known only to native peoples in Mexico. Moreover, this investigation had arisen out of his curiosity about psychedelic plants, which derived, in turn, from his synthesis of the world's first man-made magic substance, LSD.

Hofmann is not known to have been particularly mystic, but he nonetheless invoked a vivid image to describe this phenomenon. He wrote of "an unusual cycle of chemical research filled with coincidences, a kind of magic circle, which started with the synthesis of various lysergic acid amides and the discovery of the extraordinary psychotomimetic potency of lysergic acid diethylamide (LSD) which led to the investigation of the sacred Mexican mushrooms, the isolation of psilocybin, and ended with ololiuqui, where lysergic acid amides were again encountered, thus closing the magic circle."

Hofmann's metaphor of a magic circle is particularly apt. Essentially he accomplished the alchemical goal of transforming matter into some form of psychic energy. Adding two ethyl groups to lysergic acid amide not only increased its potency by twenty to forty times, it subtly transformed the nature of the psychic effects: An apathetic indifference and fatigue yielded to feelings of mystical union, cosmic connectedness, and transcendence of the usual boundaries of time and place.

That a chemical could induce such changes—and in such minute doses—suggests that hallucinogens and psychoactive chemicals are structurally related to specific neurotransmitters in the brain. Specifically, mescaline is structurally similar to epinephrine and norepinephrine, and LSD to serotonin. Psilocybin and psilocin—the active ingredients of hallucinogenic mushrooms—are tryptamine derivatives, as is the neurotransmitter serotonin. These structural similarities make it highly probable that the effects of hallucinogens and psychedelics are the result of interactions with receptor sites within the brain. Moreover, these interactions are exquisitely sensitive to even small changes in the spatial arrangement of the atoms in the molecules. Thus, a dose of .05 milligram of LSD will induce hallucinations that last for several hours, while a form identical to LSD except for the spatial arrangement of the atoms will be psychically inactive at doses ten times greater.

Perhaps the most valuable and important contribution of Hofmann's research concerned mental illness. As he himself pointed out, rather than demons or dybbuks or, in more modern times, psychological conflicts, neuroscientists now had reason to believe that emotional illnesses might be the result of biochemical changes taking place within the brain. This research has relevance for our understanding of the normal mind and brain as well. The work of Hofmann and others points not to a separation but to an association between the mind and the brain. Our minds do not exist in a vacuum; they are linked to the chemistry of the brain.

"Neither thought nor emotion can occur without some chemical change," wrote Dr. Robert S. De Ropp in his influential 1957 book, *Drugs and the Mind.*

> The cruelty of the tyrant, the compassion of the saint, the ardor of lovers, the hatred of foes all are based on chemical processes. However hard we may try, however earnestly we may wish to do so, we cannot separate mind from matter or isolate what we call man's soul from his body. Were this not so the action of drugs on the mind could never be understood. It is precisely because all mental and emotional processes have a chemical basis that these drugs exert an action. If mind existed in a vacuum apart from matter, we would not be able to influence it by drugs.

Unfortunately, recent events have been far from encouraging in regard to the espousal of a chemical approach to inner transformation. To this day, one of Albert Hofmann's deepest sorrows has been the destructive effects that his brainchild, LSD, wrought on a society that fundamentally misunderstood the implications of his work. Coming as it did at a time of great social upheaval— the Vietnam war, sexual liberation, the civil rights movement— LSD provided a convenient means of withdrawal rather than engagement. "Turn on, tune in, drop out," Timothy Leary, the guru of the psychedelic movement, put it in the mid-1960s. Only a few years after advocating this approach to the chemical modification of the brain, Leary was in jail, and the psychedelic movement no more than a bad dream that everyone wished to put behind them. But LSD was not the end of mankind's efforts to achieve self-transcendence through chemistry; nor is it likely that this search will ever end.

"Our nervous system . . . is arranged to respond to chemical intoxicants in much the same way it responds to rewards of food, drink, and sex. Throughout our entire history as a species, intoxication has functioned like the basic drives of hunger, thirst, or sex, sometimes overshadowing all other activities: Intoxication is the fourth drive. Individual and group survival depends on the ability to understand and control this basic motivation," writes psychopharmacologist Ronald Siegel in *Intoxication: Life in Pursuit of Artificial Paradise.*

It was Albert Hofmann who started us on the path toward understanding and controlling this natural bent and wise use of chemicals to modify internal processes. His research provided the impetus for three decades of gathering of new knowledge about brain receptors. Nor was he a solitary laborer in this field. Other investigators were working at the same time who were also opening up new vistas on the brain at the chemical and molecular level.

CHAPTER 5
LETHARGY IN A GUINEA PIG

*I*n the late 1940s an Australian psychiatrist formulated a hypothesis about manic depression, an illness that we now call bipolar disorder in which periods of depression alternate with episodes of mania. Working out of a "small chronic hospital with no research training, primitive techniques, and negligible equipment," as he described it, John Cade concentrated on devising a way to test for the presence of "toxins" circulating in the bodies of patients suffering from the manic phase of bipolar disorder.

Cade thought of mania as resulting from a "state of intoxication by a novel product of the body circulating in excess, whilst melancholia is the corresponding deprivative condition." In his search for the elusive "toxin," Cade hit upon the idea of injecting the urine from manic patients into guinea pigs. The crudity of this approach—urine contains many chemical constituents—was justified, Cade claimed. "Because I did not know what the substance might be, still less anything of its pharmacology for lower animals, the plan seemed to be to spread the net as wide as possible and use the crudest form of biological test in a preliminary investigation."

Initial results seemed to confirm Cade's hunch. While concentrated urine from any source could kill a guinea pig in sufficient quantities, the urine from a manic patient killed more quickly and at about one-third the dose. Further, the same agent was likely at work in all cases since the mode of death was invariable. After twelve to twenty-eight minutes, during which the animal appeared perfectly normal, tremulousness, clumsiness, and paralysis set in. Then the animal started twitching, epileptic convulsions followed, and ultimately death ensued.

Next Cade separated urine into its chemical components. His reasoning was elegant: "to identify the quantitative modifiers that made some specimens of urine from manics so much more toxic than any specimen from other sources." The toxic agent turned out to be urea. When urea from manic patients was injected into the guinea pigs, the mode of death was exactly the same as with whole urine. In contrast, the other two major components of urine—uric acid and creatinine—produced no toxic effects at all, even when injected in maximal doses. In fact, it turned out that creatinine is not only nontoxic, it exerts a protective action against the effects of urea. Uric acid, on the other hand, mildly enhances urea's toxic action.

In order to study this further, Cade had to overcome a major hurdle. Uric acid crystals are insoluble in water, and since he wished to "estimate more accurately how much uric acid increased the toxicity of urea," some other means of dissolving uric acid was required.

At this point he carried out an experiment that would change the lives of hundreds of thousands of people suffering from manic-depressive illness. "The practical difficulty was the comparative insolubility of uric acid in water," he wrote later, "so the most soluble urate was chosen—the lithium salt."

Cade injected several guinea pigs with urea saturated with lithium urate, then sat back to watch for the urea's toxic action to be enhanced. Instead, just the opposite happened: Urea's toxicity was *reduced,* not enhanced. Intrigued, Cade shifted his attention from urate to lithium itself. "It appeared as if the lithium ion may have been exerting a protective effect."

This was a bizarre idea, since lithium is present in the universe in only small quantities in comparison with its chemical relatives, sodium, potassium, magnesium, and calcium. From the viewpoint of probability, lithium seemed an unlikely candidate for altering the toxicity of the agent or agents in the blood of manic patients that was responsible for their mood alteration.

To investigate further the effects of the lithium ion, Cade switched from lithium urate to lithium carbonate, which is better tolerated and less likely to cause troublesome side effects. He found that the mortality of guinea pigs injected with urea decreased by fifty percent if lithium carbonate were added to the solution. At this point he abandoned all interest in urea and uric acid in order to concentrate on lithium itself.

When he injected a large dose of pure lithium carbonate into his guinea pigs, Cade was startled to observe "that after the injection they could be turned on their backs and that, instead of their usual frantic . . . behavior, they merely lay there and gazed placidly back at the experimenter."

Cade continued: "It may seem a long way from lethargy in a guinea pig to the control of manic excitement, but as these investigations had commenced in an attempt to demonstrate some possibly excreted toxin in the urine of manic patients, the association of ideas is explicable." He decided to try lithium carbonate as a treatment for mania, but first he had to make absolutely certain that he was in compliance with the oldest maxim in medicine, *Primum non nocere:* "First, do no harm." The first human subject to swallow the lithium carbonate pills was none other than Cade himself. When no harmful effects occurred, he administered the pills to ten manic patients.

The first patient was a fifty-one-year-old man, "wizened . . . and in a state of chronic manic excitement for five years." After five days on lithium, the calming action of the drug was evident: The man was more settled, tidier, less uninhibited, and less distractible.

Over the next two months the man improved to the point that he was ready for discharge, a development so unexpected that he had to be kept in the hospital for another two months in order to accustom him to the prospect of freedom: "As he had been ill so long and confined to a chronic ward, he found normal surroundings and liberty of movement strange at first." Finally, four months after starting the lithium, the patient was released and told to continue the drug.

Over the next six months, however, he became increasingly careless and forgetful about taking his medicine. Such a turn of events is not unusual; euphoria and the sense that one is invincible are two of the core symptoms of mania. Too often, the patient stops the medication, and the spiral of manic excitement, disorganization, and rehospitalization begins once again. In this instance, irritability and erratic behavior forced the patient's return to the hospital. But a month after the reinstitution of lithium, he was "completely well and ready to return to home and work."

Cade's results with nine other manic patients were "equally gratifying." Excitement, euphoria, restlessness, irritability—these symptoms disappeared or were greatly improved, usually within a week or so after starting on lithium.

In September 1949, Cade published his findings in *The Medical Journal of Australia.* His "Lithium Salts in the Treatment of Psychotic Excitement" is a remarkable paper for several reasons. First, the experiment, if judged by Cade's stated purpose "to demonstrate some possible excitatory toxins in the urine of manic patients," must be deemed a failure. Neither he nor anyone after him has ever convincingly demonstrated the existence of any "toxins" in the urine or anywhere else in the body that are responsible for the symptoms of mania.

Second, the discovery of lithium's effectiveness came as a surprise—what scientists refer to as a serendipitous finding. It was as if Cade were searching for something in his lab, stubbed his toe, and only then discovered the object of his search lying right there at his feet. But this does not mean that his success was simple luck. On the contrary, his experiment was a classic exercise of cool logic, accompanied by the experimenter's willingness to modify initial impressions and hypotheses on the basis of what is encountered and observed.

Finally—and this is something that we will encounter many times in our exploration of how the brain is modified by chemicals—current knowledge of the brain played no part in his discovery of the effects of lithium. Cade had started with a hunch that in fact was a modernized version of Hippocrates's two-thousand-year-old humoral theory.

Hippocrates and the school of medicine that he founded in 400 B.C. proposed that mental illnesses are the expression of imbalances among four bodily fluids, or humors—blood, which makes people sanguine; phlegm; yellow bile, known as choler; and the black bile of melancholy. These are normally present in harmonious proportions, said Hippocrates, and each of them is associated with a specific temperament that, if greatly exaggerated, develops into an identifiable mental illness. Thus, a sanguine temperament is marked by liveliness; in excess, that same sanguine disposition is transformed into mania.

If we substitute the word *humor* for *toxin,* Cade's hypothesis concerning mania would have made a good deal of sense to Hippocrates. Indeed, Cade's procedure required little scientific sophistication. It wasn't necessary to know anything about neurotransmitters or receptors; nothing in the way of specialized equipment was required; neurochemical analysis played only a small part; and the part it did play was at the level of a medical school biochemistry course.

Cade's achievement should not be undervalued—far from it; he was innovative, logical, and courageous. Nonetheless, one of the most important discoveries about mental illness and the brain did not evolve from an elaborate or highly sophisticated theory.

If uric acid had happened to be more soluble in water, he would never have taken lithium in the form of lithium urate off the shelf. Would someone else have discovered lithium's antimanic activity? It seems unlikely. Even now, forty years after Cade's experiment, lithium's exact mechanism of action remains unknown. But one thing is known with certainty: Mania does not result from a lithium deficiency in the body. In other words, there is no scientific reason why anyone would have deliberately selected lithium to try as an experimental treatment for mania.

"It is somehow surprising and fascinating," wrote neurobiologist Daniel C. Tosteson, "that a simple salt, an ion, an extract of rock, is able to alter such an ephemeral and subtle property of mind as mood." Tosteson wrote this in a 1981 article in *Scientific American,* at a time when understanding of the brain was burgeoning. Imagine just how improbable Cade's hypothesis must have seemed when he made it more than thirty years earlier. On our journey into the world of brain functioning on the chemical and molecular scale, we will not encounter anything quite like this again. All mind-altering drugs that have been discovered or synthesized since Cade's day are complex substances that are often transformed within the brain and elsewhere in the body into active forms that can exert powerful behavioral effects. Indeed, given lithium's physical-chemical simplicity, its effectiveness was so unexpected and anomalous that for the longest time, suspicions were raised about its being a placebo effect. However, comparison studies show that not only is lithium superior to a placebo, it is helpful in controlling mania over eighty percent of the time. It also has a favorable effect on depression. In economic terms, lithium treatment has been conservatively estimated to save over $400 million per year in reduced medical costs and increased productivity in persons with this disorder.

Every year, on the day that the Nobel prizes are announced, I think of Cade. If ever there was a scientist who deserves such recognition, it is John Cade. His discovery of the antimanic effects

of lithium inaugurated the modern science of psychopharmacology.

Twenty years ago, at an international meeting near Baltimore
to honor the discoverers who changed the face of psychiatry in
the twentieth century, I met John Cade. A tall, thin, and elegant
man, he was clearly an outsider: He wasn't jetting back from yet
another brain meeting in Vienna or the Vatican or Brussels. Like
myself, he was a practitioner, not a researcher, a man for whom
the word *patient* meant a real-life face and name, not just a number on a research proposal.

It was obvious from the way others approached and talked to
him that Cade was also something of a legend. Encountering him
among these academics was something like meeting a person who
has just returned from a mythical kingdom or to employ a more
contemporary analogy, encountering a man who in his first major
fight had knocked out a great champion and then promptly re-
tired, never to fight again. How had he done it? Could he do it
again? The doubters and the ponderers would never know. Like
a mountain or a monument that indisputably exists, Cade simply
stood there surveying the large audience, a living testimonial to
the fact that genius may sometimes speak only once, provide only
a single utterance against oblivion. Behind him I could see in my
mind's eye that seemingly endless column of patients suffering
from manic depression whose lives, thanks to Cade, were better
now.

CHAPTER 6
CLASPING THE NEURON

At about the same time as John Cade was discovering the antimanic effect of lithium, a French surgeon named Henri-Marie Laborit was attempting to develop a drug capable of counteracting surgical shock. Laborit's service as an army surgeon during World War II had convinced him that despite remarkable advances in surgical techniques, shock remained the greatest challenge to further progress in his chosen field.

The term *shock* has varied meanings, many of them casual and imprecise, but in medicine it refers to a specific bodily response. Patients who are in shock are generally apathetic, weak, and pallid; their faces are sunken; their expressions are of unmitigated anxiety and dread. Their skin is moist, cold, and gray; their pulse is rapid and only faintly palpable; their blood pressure is low; their breathing is shallow and labored. If capable of speech, they express foreboding and fear.

This array of physical signs does not mean that shock occurs only as a result of physical injury. Rigid distinctions between the physical and the mental are not only unhelpful here but can be misleading. Some people, for example, who learn of the death of a friend in an automobile accident, suffer a mental event that has all the physical signs of shock. But Laborit's initial investigations were not concerned with emotional causes; he was seeking ways to favorably influence those forms of shock that surgeons encounter.

Like other surgeons at the time, Laborit was convinced shock resulted from the effects of a naturally occurring chemical, histamine, on the brain—specifically, on the autonomic nervous system, which influences pulse, blood pressure, respiration, and other physical processes implicated in shock.

Antihistamine drugs had long been employed the world over as medications for hay fever, asthma, and other allergic reactions. Like most agents that affect the nervous system, antihistamines exert both peripheral and central effects. The peripheral effects reduce sneezing, nasal congestion, and itching of the eyelids—a therapeutic triumph. But the effects of most antihistamines on the central nervous system are undesirable; the sleepiness, light-headedness, and sedation can seem more intolerable than the allergies themselves on occasion. In an attempt to correct this situation, pharmaceutical houses were vying with one another to develop an antihistamine that produced marked peripheral effects but no significant central effects—a goal, incidentally, that despite forty years of research has still not been completely achieved.

But Laborit's aim was just the opposite: Rather than less sedation, he was seeking a drug that would induce greater central effects, especially a calming, relaxing, though not necessarily sedative action. As he put it, Laborit wanted to discover or synthesize a compound that could reproduce in humans the ataraxia extolled by the Stoics in ancient Greece.

The term *ataraxia* ("imperturbability") was coined by Howard Fabing, a neurologist, and Alister Cameron, a professor of classics at the University of Cincinnati. In Greek, *ataraktos* means "undisturbed" and comes from the verb *ataraktein,* "to keep calm."

In addition to its euphonious sound, the word *ataraxic* conveys a nuance that is absent from descriptions of drugs as sedatives or tranquilizers. (The term *tranquilizer* dates back to 1810, when Benjamin Rush, the founder of American psychiatry and a signer of the U.S. Constitution, designed a wooden chair, equipped with restraints, that he called the tranquilizer.) For the Greeks, *ataraxia* implied steadiness, inward calm, and harmony; it could be achieved through two channels: religion and philosophy. Both approaches require long and arduous self-discipline. It is not surprising, therefore, to find speculation throughout the centuries about possible ways to achieve a shortcut to ataraxia through chemical means.

An ataraxic chemical would be one that brought about mental balance by correcting disturbed states of mind; it would reestablish mental equilibrium either by calming the agitated or stimulating the withdrawn and inactive.

In search of an ataraxic, Laborit worked with a compound he obtained from the pharmaceutical firm Rhône-Poulenc in Paris in June 1951. The compound was identified simply as 4560 RP. Chemically related to a popular antihistamine, 4560 RP exerted a powerful central nervous system effect on test animals. When Laborit administered it to patients, he discovered that it lessened presurgical anxiety and normalized the stress reactions of the autonomic nervous system, earning it the title "autonomic stabilizer." Patients did not lose consciousness; they simply exhibited a mild drowsiness and a detachment from the preoperative preparations going on around them.

Nothing like this had ever been previously encountered. Preoperative sedation by drugs had always been achieved at the expense of alertness: The perfectly sedated patient was one step removed from one in an anesthetically induced coma. But the central nervous system effect of 4560 RP was unique. The patients remained awake if not entirely alert, they responded to questions, and they voiced fewer fears and anxieties. "There is not any loss in consciousness, not any change in the patient's mentality but a slight tendency to sleep, and above all disinterest for all that goes on around him," Laborit wrote in his journal.

At this point in 1951, this is how things stood: A thoughtful and ingenious surgeon and an innovative drug firm interested in antihistamines that had greater rather than lesser effects on the brain had successfully collaborated in producing a drug capable of inducing calm and relaxation without significant sedation in patients both before and after major operations.

The discovery of a drug capable of stabilizing the autonomic nervous system and thereby combating shock, Laborit's original objective, was an advance of undeniable importance in surgical specialties. But its implications for the treatment of mental illness were not immediately evident. Someone had yet to suggest that surgery could teach something to psychiatry: that an agitated and distressed patient on a psychiatric ward might respond to a drug with proven effectiveness in producing ataraxia on the surgical ward.

Laborit took up that challenge by encouraging a few psychiatrists he knew personally to administer 4560 RP—eventually named chlorpromazine—to their restless patients. On January 19, 1952, in a mental hospital in the town of Montauban, near Toulouse, Jacques, a twenty-four-year-old sufferer from recurrent mania, was given the drug. Electroshock treatment and restraints had proven only minimally helpful for him in the past; Jacques had experienced the return of his symptoms and had been hospitalized three times. But this time things were different. After twenty days on chlorpromazine, Jacques was judged sufficiently improved to be discharged and resume his life outside the mental hospital. Other patients tried the drug in mental hospitals in France and Italy; one aggressive paranoid patient was rendered "charming and docile" after only three days of treatment, it was reported to the Société Médico-Psychologique in Paris.

Success stories accumulated rapidly from that point on, but no one had any idea how the drug actually worked. Still, certain observations could serve as guidelines for finding out. As a general rule, biological treatments for emotional and mental illnesses bring about alterations, even disturbances, in normal brain functioning. Chlorpromazine and drugs like it induce chemical changes in the brain. At higher doses they can generate symptoms typical of Parkinson's disease: The patient is rigid, moves with great slowness and difficulty, and displays tremors of the hands and often the face. Abrupt spasms of the neck (torticollis) can suddenly occur, as can abnormal eye movements in which the patient seems compelled to gaze upward. "It was found that all drugs producing these characteristic neurological syndromes had similar antipsychotic activity," recalls Jean Deniker, who at Laborit's suggestion had carried out (with Jean Delay) the original experiments on chlorpromazine in acutely psychotic patients. "In contrast, compounds of the same chemical group that did not cause neurological effects had almost no therapeutic activity in psychosis."

Deniker's observation that certain forms of psychosis, especially schizophrenia, could only be improved by medications that exerted adverse neurological effects suggested that an antipsychotic drug's effectiveness was related to its action on the brain. (Indeed, the term applied to antipsychotics, *neuroleptic,* means "to clasp the neuron.") Further, since Parkinson's disease in its natural form was known to result from a disturbance in brain chemistry in the substantia nigra, a subcortical brain area, it followed that neuroleptic-induced Parkinson's disease was most likely to be the result of a drug's effects on the chemistry of this and perhaps other subcortical brain areas.

Psychiatrists were now in possession of a powerful new drug that could control although not cure certain forms of psychosis. Thanks to chlorpromazine and other antipsychotic agents developed in the same period, neuroscientists were learning that mental illness could not be meaningfully understood without reference to brain functioning.

From the vantage point of today, it is difficult to convey just how revolutionary this thinking really was at the time. Even as recently as the late 1960s, despite insights like those of Albert

Hofmann a quarter of a century earlier, psychiatry—certainly American psychiatry—was heavily committed to the idea that mental illness resulted from psychological causes. Those few psychiatrists and psychiatric trainees who suggested that much mental illness might have physical causes—and I include myself here—were invited to pursue other specialty interests. (I switched to neurology, while others went into neurosurgery or the more basic science of neurophysiology.) This is particularly ironic considering that lithium and chlorpromazine were developed as treatments in the late 1940s and early 1950s. In other words, by that time all the indicators were available for the disinterested and objective observer to conclude that the mysteries of the human mind could be revealed only by studying and understanding the brain—particularly the effects of chemicals on brain function. But from this point onward the pace would quicken. Within a decade the importance of neurological factors in the causation and treatment of mental illness would no longer be in doubt.

CHAPTER 7
THE ANATOMY OF MELANCHOLY

*A*lthough chlorpromazine provided the most dramatic innovation in the treatment of psychosis, neither it nor lithium was the first substance known to exert a powerful effect on patients with mental illness.

In 1931 two researchers in India had published a paper in the obscure journal *Indian Medical World* in which they spoke of "a new Indian drug for insanity." The drug, extracted from the root of a plant known as *Rauwolfia serpentina,* was actually not new in the sense of being recently discovered. It had been used for thousands of years in India; ancient Indian texts described anxious, agitated, and overwrought patients who were calmed under its influence.

Over the next eighteen years other papers from India confirmed the antipsychotic effect of rauwolfia. But at that time, unlike today, news of scientific breakthroughs traveled slowly, particularly from an area of the world not known to be on the cutting edge of brain research. New research findings and hunches about possible rewarding avenues of inquiry largely depended upon word of mouth. Sometimes a researcher would stumble upon an interesting line of inquiry while idly paging through a journal during a laboratory break. On other occasions the impetus would come when a visiting graduate student casually mentioned an observation he had made at a medical center halfway around the world. We do not know how knowledge of rauwolfia's usefulness spread, but eventually it found its way to Switzerland and the attention of a research team at Ciba laboratories headed by pharmacologist Hugo Bein.

Bein set out to discover the active ingredient of rauwolfia, but this promised to be far from easy. The plant contains dozens of active substances. Compounds of different pharmacological activity could be extracted from it depending on the chemical processes used. In addition, in certain derivatives the tranquilizing effect was offset by other components that produced a stimulating effect. At this point, Bein turned to a medically traditional but in our age controversial avenue of approach: "It has always been my opinion that in the case of a centrally acting substance, the pattern of effects produced in the unanesthetized laboratory animal is of primary interest," says Bein.

By placing extracts of rauwolfia in the eyes of hares, Bein made a crucial observation: Contraction of the pupils correlated with tranquilization. Guided by this link, Bein and his associates isolated reserpine, the active component of rauwolfia.

It is fascinating to consider the implications for our understanding of the human brain if knowledge of rauwolfia had spread beyond India in the early 1930s instead of the early 1950s. Certainly a treatment for the more agitated forms of psychosis would have been available twenty years earlier. This indisputable benefit might have been offset, however, by the probability that given reserpine, Laborit might not have made his concentrated and grimly determined effort to convince his psychiatric colleagues to try chlorpromazine.

As it turned out, chlorpromazine quickly established itself as far superior to reserpine in the treatment of severe mental illness. Reserpine has a weaker antipsychotic effect that is slow in onset; even at the optimum dose improvement may not take place for two months or more. Reserpine research, however, was far from a blind alley. It would go on to exert a powerful and unexpected influence on another psychic ill that besets humankind: depression.

Throughout recorded history two competing concepts of depression, or melancholy, vied for attention. The Hippocratic theory considered depression a physical disease caused by the humor black bile. Curing melancholy consisted of administering to the patient a medication with the power to neutralize this humor.

Vying with this theory was the notion that melancholy represented the reaction on the part of a *susceptible* person to stressful life events. In this view only a person with a melancholic disposition would allow losses or disappointments to depress him more than briefly.

Not surprisingly, the physical theory was most appealing to physicians, at least until the advent of psychoanalysis. The theory was bolstered by the belief that the brain was the organ responsible for melancholy; Hippocrates had said, "And by the brain we become mad and delirious, fears and terrors assail us, some by night and some by day . . . all these things we endure from the brain when it is not healthy."

Proponents of the alternative "melancholic temperament" theory included poets and writers such as Robert Burton. His major work, *The Anatomy of Melancholy,* is very definitely tilted toward the notion of melancholy as a personality trait rather than the result of a physiological source that in theory could affect anyone. Three centuries later, Freud's thinking, as set forth in *Mourning and Melancholia,* was very much in this tradition, too, emphasizing the "psychogenesis" of depression.

From this conflict between the advocates of the biological and the melancholic theories emerged a compromise: a distinction between *endogenous* or biologically induced depression, and *exoge-*

nous or reactive forms. Depending on one's orientation, melancholia might be considered a not entirely undesirable character trait consisting of a somber seriousness and a tendency toward introspection. According to this view, melancholia could get out of hand when the melancholic person retreated from contact with others, or perhaps even moved to the extreme of taking his own life, but there was nothing specifically neurological or chemical about such unfortunate developments.

Alternatively, endogenous melancholia could be the behavioral expression of a chemical imbalance in the body, principally in the brain. Its cure would consist in restoring that balance.

In the 1950s several types of evidence were available to the keen observer and careful thinker that favored the biological theory. Methamphetamine ("speed"), which had been synthesized in the 1930s, prolonged wakefulness, increased alertness, and raised the spirits—sometimes to the level of a "high." These temporary improvements occurred in persons of different temperaments even when there was no change in their life circumstances; but among those who were depressed, the depression returned with renewed vigor whenever the drug was discontinued, thus leaving the patient worse off than before taking amphetamines.

Another clue pointing to the brain came from electroconvulsive therapy (ECT). The first deliberately induced seizure for the purpose of helping a patient with depression had taken place in 1934, when a Hungarian psychiatrist, Ludwig von Meduna, used intravenous injections of camphor oil to provoke it. What reasoning provided the basis for this radical treatment? Nothing more substantial than an observation in the previous century that schizophrenics who also happened to suffer epileptic seizures sometimes showed improvement in their schizophrenia after a spontaneous seizure. If a spontaneous seizure could effect improvement, the reasoning went, then a deliberately provoked one should do so equally well.

A few years later, two Italian psychiatrists, Ugo Cerletti and

Lucio Bini, substituted electrically induced seizures for camphor oil injections. The treatment was found to work for carefully selected patients suffering from depression. The only other treatments for depression were in essence variations on the "melancholic temperament" theory, the various forms of psychotherapy that were then favored by the majority of psychiatrists. But in 1956 treatment approaches for depression took a very different turn. The impetus for this change was reserpine.

Rauwolfia enjoyed a reputation in the East for effectiveness not only in the treatment of psychosis but in the control of high blood pressure. In about ten percent of patients who took it for blood pressure, rauwolfia, especially its active ingredient reserpine, also induced depression. Reserpine-treated animals (usually rats or mice) eventually became laboratory models for depressed humans because they exhibited many similarities in behavior, including reduced movement, the so-called psychomotor retardation characteristic of human depression. Research was focused on one question: Could a drug be found or developed that reversed reserpine-induced depressionlike behavior in rats or mice? If so, would that drug be useful in the treatment of depression in humans?

The answer to those two questions came from another source, also medical in origin but hardly psychiatric: the treatment of tuberculosis.

Until about 1955 tuberculosis was a major chronic illness that cried out for definitive treatment. In 1952 a new drug, isoniazid, was a promising candidate for the much-sought-after cure. Many patients treated with isoniazid also experienced a distinct elevation in mood. Not surprisingly, most of the doctors treating them attributed this rise in mood to the general improvement in their health and well-being that accompanied recovery from a serious illness. But some clues suggested that the mood improvement should not be interpreted quite so facilely. Not all patients experienced mood elevations, and in those who did the change sometimes went far beyond a mere lifting of spirits—on occasion it involved an actual toxic psychosis marked by overexcitement. No one had an explanation for this.

On April 11, 1956, Nathan Kline, a brilliant and flamboyant psychiatrist with a penchant for using daring treatment approaches to mental illness, delivered a lecture at Warner Laboratories in Morris Plains, New Jersey. Afterward, researcher Charles Scott came up to talk to him. Caught up in that mix of fatigue and elation that follows public speaking, Kline was more in a mood to talk than to listen, but what Scott said immediately captured his attention.

Scott told Kline of his observation that if rats were given isoniazid *prior* to reserpine, the usual sedation and calming actions of reserpine were replaced by hyperactivity and hyperalertness. Kline immediately realized that this was why some patients given isoniazid for tuberculosis experienced a mood elevation. He was inspired to try a novel and intriguing approach to depression. "The possibility of using this combination on depressed patients immediately led me to speculate whether this was the psychic energizer for which we had all been looking," he later wrote.

At the beginning of November 1956, Kline started this isoniazid-reserpine combination in seventeen seriously ill hospital patients who had been diagnosed as suffering from schizophrenia, and in nine depressed patients drawn from his private practice. In his 1958 report on the experiment, Kline claimed a 70 to 86 percent favorable response in severely depressed patients. This success figure was considerably above anything reported before or since for depressed patients.

Along with the alleged favorable response in the patients came various explanations of isoniazid's effectiveness. One bizarre theory—a measure of the ignorance of drug action at the time—held that the antidepressant action of isoniazid was due to the fact that it shared a common chemical derivative with a rocket fuel that was then in use!

Another explanation focused on the drug's action as an inhibitor of a brain enzyme called monoamine oxidase (MAO). After a presynaptic neuron releases the monoamine neurotransmitters dopamine, serotonin, and norepinephrine, they travel across the synaptic cleft and attach to their specialized receptors on the post-synaptic membrane. They are then rereleased back into the synapse. There they are inactivated in one of two ways. Either they are reabsorbed by the presynaptic neuron in a process called reuptake, or they are broken down within the presynaptic neuron by the enzyme MAO. If MAO is prevented from acting by means of a MAO reuptake blocker, the neurotransmitters remain intact within the synaptic cleft, and they continue to exert their effects for longer periods of time. Isoniazid is such a blocker. It inhibits the action of MAO and thereby increases the concentration of monoamines in the brain. This was hypothesized as the basis for the drug's antidepressant effect.

Support for this hypothesis came when neuroscientists discovered that reserpine's action within the brain resulted in an effect opposite that of MAO inhibitors (MAOIs). Instead of increasing the quantity of neurotransmitters in the synapse, reserpine decreased it. Instead of improving the depressed person's spirits, reserpine lowered them even more.

On the basis of these observations, a theory evolved about the origin of depression. In its simplest form this theory associated depression with an absolute or relative deficiency of certain neurotransmitters at functionally important synapses of the brain. Mania, at the other end of the spectrum, was associated with an excess of these neurotransmitters.

Isoniazid's antidepressant activities were consistent with this hypothesis. Inhibiting the action of MAO made more norepinephrine, the major neurotransmitter that was thought to be involved in depression, available within the synaptic cleft. As the norepinephrine level increased, depression lifted and the patient was restored to his or her previous health—or so the theory went.

But, as mentioned above, neurotransmitters are also inactivated by a reuptake process. Ordinarily, after release from a receptor on the membrane of a postsynaptic cell, a transmitter is taken up by the presynaptic neuron, and its action ceases. To retain the neurotransmitter within the synapse for a longer time so that it continues to act, this reuptake process can be blocked.

The discovery of the reuptake-blocking family of antidepressants evolved from a centuries-old observation about opium. In an authoritative psychiatry textbook in 1928, the psychiatrist J. Lange confirmed that melancholy and the effects of opium were similar. "Opium appears to have a specific effect on melancholy," he wrote, "admittedly only a symptomatic one, and this is particularly striking because opium itself causes the same effects that resemble the symptoms of melancholy."

Lange's statement led to a working hypothesis: If opium could help patients suffering from depression, then other drugs might work even more effectively than opium. But where should scientists look for them? For thirty years that question remained unanswerable.

Finally in the mid-1950s, in the absence of any compelling alternative, research psychiatrists investigated whether the new antipsychotic drug chlorpromazine, so effective in treating schizophrenia, might also have a beneficial effect on depression.

Roland Kuhn, a Swiss psychiatrist, tested chlorpromazine on a few depressed patients but found that though the drug did exert a calming effect, the depression itself was unaffected. In 1956, on the theory that a different drug with a similar chemical structure might be more effective, he developed a related compound, then known simply as G22355, and gave it to three patients suffering from severe depression. It soon became clear that the drug, later called imipramine, possessed strong antidepressant features. Kuhn reported these findings at the Second International Congress of Psychiatry in Zurich in September 1957. Barely a dozen people were in attendance, and his paper was received with much skepticism. But this gave way to acceptance as the drug's effectiveness was confirmed.

Kuhn's success with imipramine had depended greatly upon his skill in separating out from the mass of depressed patients only those who were likely to benefit. "The substance is particularly effective in typical endogenous depressions, provided there is a vital disturbance standing clearly in the foreground," Kuhn wrote in his report. But as he soon discovered, very few psychiatrists understood what he meant by a "vital disturbance."

Most people associate depression with tearfulness, self-accusation, outward distress, and agitation, often leading to suicidal tendencies. None of these symptoms were what Kuhn meant by a "vital disturbance": "In a vital disturbance the patients complained of lassitude, feelings of heaviness, depression, and inhibition, a slowing down and difficulty in thinking and acting, of internal tension and the inability to feel pleasure, the whole picture subject to diurnal fluctuations with exacerbation in the morning."

Notice that Kuhn's emphasis was less on an objective observation of a patient's behavior (Did she cry? Was he pacing the floor with agitation and remorse?) and more "on the subjective experience of the patient." Only in those patients who experienced the subjective components, the "vital disturbance," was the drug helpful.

With this discovery Kuhn at last did away with the ancient but artificial distinction between a depression induced by the environment (exogenous) and a depression stemming from a biological source (endogenous). What was most important in predicting a patient's response to imipramine was the *subjective* character of his or her experience: "The drug largely or completely restores what the illness has impaired—namely, the mental functions and capacity and, what is of prime importance, the power to experience."

Kuhn's findings produced a familiar situation: A highly successful treatment existed for patients properly selected, but no one knew the explanation for what was happening. Imipramine's mechanism of action, however, was discovered soon thereafter. The drug is a *tricyclic antidepressant* (a class of drugs so named because of the presence of three fused rings in their structure);

researchers discovered that these drugs actively blocked the reup-take of the neurotransmitters norepinephrine and serotonin.

Note that this biogenic amine theory emphasizes the rise and fall of neurotransmitters but places no emphasis on the actual mechanisms whereby the reuptake of the neurotransmitter may be blocked. What were these mechanisms?

Subsequent research concentrated on receptors within the membranes of the presynaptic and postsynaptic nerve cells. The tenacity of binding of a drug to a particular receptor, called af-finity, can vary tremendously. In general, the higher the affinity of a drug for a receptor, the greater the effect of the drug at a given dose. Drugs that block receptors, called antagonists, work by preventing access of a particular neurotransmitter to its receptor.

An antidepressant's ability to influence a receptor is a phar-macological action. Its mood-elevating properties are seen at the behavioral level. There is good reason to think these two actions may be quite separate. It usually takes from two to four weeks for mood alteration to occur, but the pharmacological effects are ap-parent within hours of the first dose. The patient may complain of blurred vision, dryness of the mouth, sedation, and drowsiness, all resulting principally from the antidepressant's effect on the autonomic nervous system.

Because of this two-phase response—side effects followed sometime later by antidepressant activity—neuroscientists have long sought more explicit proof of a *causal* connection between the biological action of these drugs and improvement in mood.

Several claims have been made in the past that a particular neurotransmitter was the key element in depression, but these claims have not withstood close inspection. It now seems most likely that depression is related to the malfunctioning of more than one neurotransmitter—perhaps many. This should not be sur-prising since moods involve many factors—thoughts, feelings, and a sense of physical well-being or malaise, to name just a few. Each of these subjective experiences inevitably involves different biological parts of the brain and the interplay of many, many neuro-transmitters.

A symphony cannot be experienced or explained by concentrating on one instrument, no matter how enthralling its performance may be. Nor is it likely that depression, an illness that affects our being so fundamentally, is biologically linked with alterations in only a single neurotransmitter. Perhaps what is needed is what Roland Kuhn mentioned in an interview: "the ability to invent something completely new, something hitherto unknown, namely, a new disease."

What is required is the discovery of a biological link that unites the patient's state of mind, the doctor's observations, and relevant transmitter-receptor correlations. Until such a link is found, it is unlikely that a satisfactory theory of depression will emerge. In the meantime we should neither be surprised nor dismayed by our inability to explain depression in terms of neurotransmitter-receptor interactions alone. Depression is not a single disease but a group of disorders that share common symptoms, each with its own unique biology.

The discoveries described so far occurred with almost no reference to processes going on within the brain. Neither Hofmann nor Cade referred to neurotransmitters or receptors in the formulation of their experiments on LSD and lithium. Laborit was the first to refer to the brain in planning his research, but he limited his efforts to the autonomic nervous system. The psychiatrists who tried chlorpromazine as a tranquilizing agent noted that the drug induced neurological symptoms referable to the substantia nigra, but this was still a great distance from explaining the drug's effects through modification of the dopamine receptor.

But with the development of the antidepressants, psychopharmacology entered into a more rigorously scientific phase. Events were now open to scrutiny at the chemical and molecular level. Thanks to the discovery and development of MAO inhibitors, imipramine, and other reuptake blockers, fathoming the mysteries of the human mind required deepening our understanding of the brain and studying what happened at the level of the cell membrane and its receptors. It is to a more detailed examination of this research that we now turn.

PART II

CHAPTER 8
RUSSIAN DOLLS

*E*ach of us exists for others as a kind of message that must be continuously decoded. Moods, preferences, habits, past experiences, secrets, hopes for the future—when these aspects of the self are concealed, the clever decoder can often read them by analyzing facial expression, body language, manners, dress, and comportment. In a way, we are like hunters scanning the terrain for clues about our prey. The reading of another person's nonverbal communication is merely a fancy way of describing our brain's ability to decode messages. It is the supreme message decoder, the medium through which all such decoding must ultimately be filtered.

Message decoding by the brain isn't limited to social interaction. Like a series of Russian dolls or Chinese boxes nested within one another, the brain encodes and decodes information on every level, from the smile and pensive last touch between parting lovers to the equally passionate clasp of one chemical for another within the helically arranged DNA molecules that make up who we are.

Our brain's potential for receiving and sending messages is determined by its structure. As we have noted, certain areas are specialized to receive light, others sound, and still others smell and taste. If nature or circumstance disrupts any of these specialized areas, the message-encoding process takes a different direction. In the case of injury or failure of development anywhere along the visual pathway, for instance, perception must operate via alternate, less efficient pathways. A heightened sensitivity to tones of voice, slight variations in cadence and rhythm, pauses, stammerings—all must now replace visual observation and provide clues for the other's intentions. In such an instance an alteration in structure profoundly modifies how messages are encoded and decoded.

When it comes to language, any disruption in the composing of messages or their deciphering can have drastic consequences. One kind of stroke, for instance, will devastate a person's ability to speak and write sentences that make sense. Despite the person's best efforts, his sentences aren't sentences at all but only a string of words that neurolinguists irreverently refer to as a "word salad." In this form of aphasia, known as Wernicke's aphasia after its 1861 discoverer, comprehension is also affected. Although the speech is rapid and effortless and marked by a superficially normal syntactical structure, the resulting sentences are remarkably empty of content and are filled with errors in normal word usage. For example, in response to the question, "Where do you live?" one patient with Wernicke's aphasia responded, "I came there before here and returned there." Obviously, message encoding and decoding are both disturbed here: the sufferer can neither correctly decode the question put to her, nor encode and deliver an appropriate response. Such deviations are the result of alterations in brain structure, in most cases damage to the posterior and superior part of the left temporal lobe.

Another form of aphasia, Broca's aphasia, results from an alteration of brain structure in a different area, usually the motor association center in the left frontal lobe, along with the posterior portion of the third frontal gyrus. In this illness the language disturbance involves encoding and takes the form of a profound reduction in the number of words employed. It can range from almost total muteness to a slowed, deliberate speech that emphasizes simple grammatical structure without articles, adjectives, and adverbs. An example of this telegraphic speech is a victim's announcement, "Ladies, men, room," in place of the intended communication, "The ladies and gentlemen are now all invited into the dining room."

But language is only the most obvious example of information transfer. We are immersed at all times in a sea of information that must be decoded: clouds that signal a thunderstorm; the sudden erratic shift of a car toward us from a neighboring lane; the dissonance of a lipstick that clashes with the color of the dress a woman wears. In each instance a message is conveyed; it is detected by the appropriate receptors; and a response is either expressed or withheld. (We may notice the dissonance between the lipstick and the dress, but in the interest of preserving harmony we elect to say nothing about it.)

Information structuring and decoding take place at every level of the brain, from the molecular to the behavioral. At the most basic level of informational organization, the *molecular* arrangements within the DNA and neurotransmitters must be exactly right. Even a tiny variation can lead to mental or emotional illness, even death.

It is considered likely that 10^{15} bits of information are stored at the next level, the *synapses*. This enormous number is beyond the capacity of even the most advanced computer currently foreseeable to replicate. Neurons are estimated to number anywhere from 200 to 300 billion, and each neuron forms anywhere from one thousand to ten thousand synaptic connections. Neurons are arranged in *networks,* whereby neuronal communication is facilitated by repetition. Habits, for instance, operate at the network level: Each time a particular pattern of neurons "fires," it becomes that much easier for the same pathway to be activated again. Then comes the level of *maps;* the brain can be parceled into regions with certain areas separated from others, like countries on the map of a continent. At the turn of the century neuroscientists competed with one another in their efforts to fashion the most complete and accurate map of the brain. Like the fifteenth- and sixteenth-century explorers of the earth's surface, these cartographers of the terra incognita of the brain operated on the assumption that it is possible to plant a flag at, say, the visual area and proclaim: "Here is the area that enables the brain to see. By searching here, we will learn to solve the mysteries of vision."

But subsequent brain researchers have discovered that such lines of demarcation dissolve when one turns from the map and encounters the land at first hand. Even something as seemingly straightforward as vision—to say nothing of such elusive processes as thinking—depends on the activation of millions of brain cells spread along a path from the retina to the cerebral cortex. Each time we speak of millions of neurons, we must remember those ten thousand synaptic connections made by some of these neurons. This pushes the informational content into figures that are virtually incalculable.

Nor is that the end of the informational levels. Over the past twenty years, maps have given way to *systems.* Neuroscientists started telling one another, "Let's not talk about where on the map of the brain vision can be located. Let's look instead at the visual system as a whole." But this focus on a more inclusive order of organization has been no more successful in helping us understand the informational content of the brain. This is because the brain itself, as a general rule, operates within certain constraints.

At each of these levels—molecules, synapses, neurons, networks, maps, systems, and the brain as a whole—information encoding and decoding reflect information from all the other levels. As your eyes move along the lines of this page, for instance, photons stimulate receptors in the retina that bring about fluctuations in the movement of ions across the membranes of those receptors. Impulses are transported along the optic nerve to the lateral geniculate, a way station on the journey to the occipital cortex. At this point billions of other neurons come into play, involving language, memory, imagination, curiosity, pleasure, and so on. As you read these lines, information is both stored and retrieved at all seven levels of organization in you. Your understanding of the words on this page is related not only to your understanding of written English but also to the chloride ion channel activity on the membranes of millions of your brain cells.

But sometimes the information encoded at various levels is not translatable into other levels. When that happens, how can these separate spheres of information storage be meaningfully related to one another? With that question we've stubbed our toe on the most fundamental conundrum about the brain: Can thoughts and emotions be translated into the language of neuronal circuits? How we answer that question depends very much on how particular we are about what we will accept as a correlation. Research on the limbic system over the last fifty years points to a strong correlation between specific structures within the limbic system and emotions such as fear and rage. A successful correlation has been achieved at the levels of systems, maps, and networks, when activating limbic structures evokes fear and rage. But if one delves deeper into the realm of synapses and molecules, the correlations are very weak and conjectural. For instance, no particular neurotransmitter or receptor interaction corresponds to fear, rage, or other emotions. Although neuropsychiatrists are prone to ask themselves whether a disordered thought is also a disordered molecule, there is an even more profound question to raise: What would it really mean if it were? Would we say to patients, "Here is a model of what's happening within your brain at the level of the nerve cell membrane when you feel depressed"? Initially this might sound impressive, but it would not bridge the gaps between a feeling of depression, a subjective event, and an intellectual understanding of a molecular event some four to five levels removed from that inner experience.

To understand message encoding at the various levels of the brain, I propose we begin literally at the beginning, with an examination of the messages that are most important for all of us: the proteins, amino acids, and nucleic acids that form DNA, our genetic code—the ultimate message that is us.

✛ ✛ ✛

The main components of living matter are proteins, which have been compared to the machinery of an organism. Proteins are large molecules composed of long chains of amino acids linked together. The sequence of amino acids along the chains determines each protein's physical and biological properties, as well as the actions of the enzymes that regulate life activities. Since proteins specify form and function, they are information molecules. Their informational content is a function of the varying number and sequence of twenty different amino acids.

In an embryo, defects or abnormalities in amino acid structure or activity can produce distortions in the transfer of information that directs the formation of brain structures. Interference with this process can result in severe brain abnormalities. The best-known amino acid deficit involves phenylalanine, which manifests itself in the inborn metabolic disorder phenylketonuria (PKU). Two of the brain's most important neurotransmitters, dopamine and norepinephrine, are synthesized within the brain by a series of chemical modifications of phenylalanine. If these modifications are interfered with, a nerve poison, phenylpyruvic acid, results. Accumulation of this abnormal molecule results in severe neurological defects and an IQ usually less than 50. Also contributing to abnormal brain structure and function are the reduced levels of dopamine and norepinephrine that result from the absence of the precursor phenylalanine.

All in all, phenylketonuria is the prototype of an *informational disease:* a disturbance in the coding and decoding of messages.

The delivery of the wrong message or, as we shall see later, even the correct message at the wrong time, can produce horrendous consequences. To the parent of a child afflicted with this illness, far more is involved than a lesson in neurochemistry. It involves a visceral encounter with events transpiring at a level far below the power of vision or, until recently, even imagination. Picture a mother sitting at home on a quiet afternoon and looking at her vacant-eyed phenylketonuric child. As she fantasizes about how things could have turned out differently, she personally and painfully learns a harsh lesson about the brain: structure determines function, and function determines behavior.

Fortunately, the treatment of phenylketonuria is successful provided the diagnosis is established by screening the blood for phenylpyruvic acid and initiating treatment within the first few days after birth. If diagnosis is delayed, too many distorted brain areas will come into being. It is as if the foundations for building the most intricate structure in the universe, the human brain, were reduced to a shambles. Paralleling the error in information coding and decoding at the molecular level—the transformation of phenylalanine—is the distortion of information processing in behavior, several levels away. Again, a mirroring occurs between events at the molecular level and events at the anatomical and behavioral levels. Disturb the brain's structure at any point along the way, from a molecule to one of the brain lobes that a surgeon can nudge with a finger at an operating table, and some form of behavior will eventually be affected. Only slight variations in structure can evoke profound consequences for survival and quality of life.

Phenylalanine is only one of twenty amino acid building blocks from which the brain and everything else in the body is constructed. This multiplicity of amino acids makes possible a nearly infinite number of combinations. For a protein one hundred amino acids long, the number of different sequences possible exceeds the number of atoms in the universe. But nature isn't profligate. She relishes the simple, and employs only a very small percentage of the immense number of possible amino acid sequences. This economy is similar to the processing of language by the human brain. Not all twenty-six letters of the alphabet are employed in all possible combinations; instead, only certain combinations are used. Moreover, there are tight controls on the frequency of letters and their placement within words and sentences. It is from such constraints that the coding and decoding processes of spoken language—writing and reading— have evolved.

The sequence of amino acids is like a language and is therefore not arbitrary. But what is the controlling influence that determines a particular sequence? To answer that we must delve deeply into the interior of the neuron, where we encounter the second category of informational molecules: the nucleic acids.

All of heredity depends upon this second category. Although other molecular species are involved in heredity—for instance, carbohydrates and lipids—they are bit players when compared with proteins and nucleic acids. The information that the nucleic acids convey ranges in complexity from fairly straightforward matters like eye color and height to far more complicated qualities like intelligence and temperament, yet their structure is intriguingly simple.

The material of the genes that bear our genetic heritage is deoxyribonucleic acid, or DNA. It carries its messages in sequences of four chemical bases, the nucleotides, which are referred to by the letters A, T, G, and C, plus the sugar deoxyribose.

The DNA molecules are built up of two chains of nucleotides entwined in opposite directions to form a helix. Each nucleotide in a chain has a complementary nucleotide in the other. They bind together in a pattern in which adenine (A) always couples with thymine (T), and guanine (G) always seeks out cytosine (C). The sequence of these base pairs and their three-dimensional conformation along the double helix determines the hereditary information.

How is this information transferred from DNA? How does a

particular sequence of bases translate into your blue eyes and my brown hair?

As a first step, the information contained in the DNA structure must be conveyed out of the nucleus of the cell and into the surrounding cytoplasm. This is the job of the second major class of nucleic acids, ribonucleic acid or RNA. While DNA is largely confined to the nucleus, RNA is largely confined to the cytoplasm.

Like DNA, RNA is composed of strings of nucleotides, but its sugar is ribose, and it contains not thymidine but the pyrimidine base uracil.

Two main types of RNA occur in the cell. The first is messenger RNA (mRNA), a single-stranded transcript of the appropriate base sequence of DNA. Then comes transfer RNA (tRNA), which is specific for each of the amino acids; thus, there are at least twenty different types of tRNA molecules. Within the cell, mRNA "copies" the sequences of bases on DNA, and tRNA acts as a go-between, linking proteins and nucleic acids. Think of the RNA molecules, both messenger and transfer, as a translation system whereby the central text contained in the DNA is transcribed and conveyed to those parts of the cell where structural elements are being put together. DNA represents the architectural plan; RNA conveys this plan to the construction crew, which then "builds to suit"—synthesizes proteins and cell structures that are specific for each of the many varied cells within the body.

Within the neuron—indeed, within all of the body's cells—the nucleic acids serve very different purposes from the proteins assembled from instructions conveyed from DNA. Nucleic acids preserve and transmit genetic information, especially information having to do with protein structure. The proteins catalyze metabolic reactions—in essence, carry on the business of living. Both functions are necessary; in fact, one isn't possible without the other. Without DNA there wouldn't be an organism, and without protein the organism wouldn't be alive, wouldn't be capable of responding to the molecular world of DNA or to the outer world of experience.

"DNA makes RNA and RNA makes protein" is how Francis Crick, the co-discoverer of the double helix structure of DNA, put it. That aphorism is, in his words, "the central dogma of molecular biology." That dogma can be roughly translated thus: The major flow of information progresses from nucleic acids to proteins, and almost never vice versa.

Moving from the molecular level to the human level and then to interactions between humans, a similar one-way flow of information occurs. Ever since the publication of Charles Darwin's *On the Origin of Species* in 1859, scientists have held that information flows from the genotype to the phenotype, from genetic endowment to individuals working out their individual destiny within a specific environment. Thus, my knowledge of psychiatry and neurology cannot be genetically transmitted to any of my children. They must learn all of it on their own. There is no inheritance of acquired characteristics or knowledge.

Notice that there is a dynamic equilibrium at work between chance and necessity, between genes and the environment. While the genetic blueprint sets forth certain possibilities—as for instance, musical accomplishment based on the inheritance of musical talents—nothing will occur if the individual grows up in a culture where music doesn't exist.

Within the brain this symbiotic relationship between genes and the environment, between heredity and experience, is particularly important. In the brain, the most biologically diverse of the body's organs, each class of cells has a distinctive form. Cells in the cerebral cortex can be distinguished from cells in the cerebellum, which in turn can be classified into many different forms. Such differences in shape and form are probably genetically determined. Mirroring these structural differences at the molecular level, mRNA exists in more than 125,000 different transcripts, which may be expressed in different brain cells and at different times.

This great variety is three to five times larger than that found in any other tissue. Such abundance suggests that more of the brain's functioning may result from genetic factors than has previously been considered likely. Not only intelligence but such personality factors as shyness, introversion, and susceptibility to certain forms of mental illness are turning out to be heavily influenced by genetics. This does not at all portend a gloomy determinism; life experiences can sometimes modify the basic personality, even transform it into its opposite. Sensitivity and the possession of rich inner resources might lead one person to write fiction or poetry, rather introverted activities; but the same personality characteristics could also result in someone overcoming shyness and taking up acting, an occupation that requires sensitivity but is usually associated with an extroverted disposition.

At the basis of individuality and diversity—that which makes me me and you you—are the various combinations of the twenty amino acids and four nucleotides that make up our proteins.

When it comes to the practical application of molecular genetics, investigations at the molecular level have turned up explanations for several diseases that involve the brain. The most clearly defined is the tragic, incurable, and ultimately fatal Huntington's chorea.

The word *chorea* comes from the Greek root meaning "dance." Sufferers from Huntington's chorea express a macabre kind of dance made up of rapid, jerky, involuntary movements that eventually involve the whole body. In most cases, however, the illness begins in small ways. An exaggerated startle response to loud sounds, irritability, mood changes, temper outbursts—these are the early indicators. But such personality changes are nonspecific, so the illness is rarely suspected until the first twitches or jerks appear, sometimes involving only the face, sometimes a series of purposeless arm or hand motions. Gradually the twisting and jerking movements spread to the entire body. Then come memory loss, confusion, hallucinations, and delusions. A deepening depression ushers in the final stages. The interval from the first symptoms until death can vary from only a few years to a decade or more.

Huntington's was found to be hereditary when geneticists discovered that practically all patients with the disease who lived on the East Coast of the United States were direct descendants of two people born in Suffolk, England, who emigrated to the United States in 1630. Several of the witches hanged at Salem are now thought to have been suffering from Huntington's. Their inexplicable movements and signs of derangement would have made them prime targets for extremists and religious fanatics, who would have sought an "explanation" for their bizarre behavior.

For over twelve generations and three hundred years, Huntington's disease has haunted the families at risk. Ultimately, the question became not whether Huntington's is inherited but how it is inherited. Family inheritance patterns provided the answer. The disease follows an autosomal dominant pattern, which means that if one parent has it, there is a fifty-fifty chance that one of the children, irrespective of gender, will inherit it. Since the average age of onset is between thirty-five and forty, most Huntington's victims already have children of their own by the time the first symptoms of the disease manifest themselves.

This combination of late onset and 50 percent transmission to children always has tragic and sometimes cruelly ironic consequences. Two brothers were under my care, whose father had died of Huntington's. One had married and fathered three children; his brother remained a bachelor. It was the married brother who in his early forties came down with the disease.

In 1983 the genetic basis for Huntington's disease was worked out, based on the demonstration of inherited variations in DNA sequences. James Gusella and his colleagues at the Massachusetts General Hospital in Boston found a consistent correlation between Huntington's disease and an abnormality on the short arm of chromosome four. After ten years of laborious and painstaking research, Gusella and his team announced the discovery of the Huntington's gene in March 1993.

Two breakthroughs, one actual and one potential, have resulted from this linkage of Huntington's disease with chromosome four.

First, the presence of the disease can now be genetically diagnosed not only before symptoms appear, but even before birth. Second, the discovery of the Huntington's gene makes possible its eventual cloning, the first step in identifying the gene product, the presence of which ultimately leads to the disease.

Consider what is involved here from the informational point of view. Defective information is contained on chromosome four. This minute component of the vast array of genetic information leads to a brain disorder marked by extreme impairments in a person's ability to move, think, and control his or her emotions. This deterioration, indeed devastation, results from a loss of brain cells through a form of genetically "programmed" cell death within the cerebral cortex, especially in the motor control centers beneath the cerebral hemispheres. In fact, this cell death is evident years before the onset of the illness, using a PET scan. Those who will go on to develop Huntington's will show less activity within the striatum, a subcortical motor area in the brain.

Reduced to a grim formula, defective DNA affects RNA affects the proteins of those elements in brain cells that affect neurotransmitter levels that affect movement (the chorea) as well as mental and emotional expression—in essence, mind. Huntington's comes very close to fulfilling the claim that a disordered thought is the expression of a disordered molecule, or in this case a gene.

Imagine yourself a neurologist or psychiatrist sitting in your consulting room on a cold and dark November afternoon. You are talking to a patient who is in the early stages of Huntington's chorea. He shows little in the way of abnormal movements, only sudden infrequent shifts on his chair—very subtle, the kind of movements you can sometimes observe, if you are very alert, in southern California during a mild earth tremor. Your patient is talking to you about his illness: his uncertainties about how long he will live, whether he wants to live, whether he wants to have his preadolescent daughter tested to learn if she too will be felled by this curse.

As he speaks you might consider for a moment, as I have on

such occasions, that you are listening to a gene talking. Despite the specifics that are being discussed, the important underlying elements are the depression, the early problems with memory, the fantasies of the horrors that lie ahead. It's as if that little twist, that distortion, in the DNA sequence of nucleotides has taken on a life of its own, a personality. As in a series of cinematographic images, your attention rapidly shifts from the level of neurons to specialized brain areas and finally to human behavior and the man himself: this specific patient talking to you this late afternoon in the gradually increasing darkness. In this series of images, matter and mind blur together.

At the level of language—your patient talking to you, and you listening, trying with all of your powers of concentration and empathy to understand him—the presence of mind is unmistakable. But mind is also discernible in his movements, in the way he frowns or smiles or adjusts his necktie. When you delve deeper into the bodily mysteries that are the man himself—peering at his PET scan, measuring his elevated levels of dopamine, isolating chromosome four from one of his body cells—such distinctions become artificial, must be done away with, since in truth, mind merges into matter. If you are not cautious, at some point you may cease to encounter the person; concepts and words may intervene. Nonetheless, something infinitely valuable has been achieved. You have discerned in the darkness and just for a moment the dreadful costs when messages carried within the molecules of the brain undergo distortion. And you have discerned the delicate skein of continuity that exists between those micromolecules and this tragic example sitting before you of what my neurology teacher so long ago and so irreverently referred to as a macromolecule—this human being.

CHAPTER 9
DOUBLE AGENTS

Sometimes message distortions within the brain are induced not by gene or protein abnormalities but by drugs or chemicals.

In 1977 a twenty-three-year-old college student in Bethesda, Maryland, came down with a sudden illness that puzzled his doctors. His hands trembled; his face was immobile and inexpressive; his body movements were rigid and painfully slow. In addition, this young man wouldn't or couldn't speak. He had a history of drug use going back nine years.

Taken to a general hospital in the area, David, as we will call him, was given an admission diagnosis of catatonic schizophrenia, a mental illness whose symptoms include extreme disturbances in movement. Typically, a catatonic schizophrenic remains in whatever position he is placed (a phenomenon known as posturing), demonstrates little spontaneous movement, and refuses to speak or otherwise communicate with those offering help.

But David was not a typical catatonic. Most catatonics don't display a tremor, as he did; their muscle tone is not rigid, as his was, but is flaccid and somewhat floppy ("waxy flexibility" is the phrase psychiatrists use to describe it). Moreover—and this was the truly disturbing part—this young man of only twenty-three looked and behaved like an elderly man. Indeed, as he lay in his hospital bed with his upper body flexed at the waist, his neck bent forward so that the chin pointed into the chest, it almost seemed as if some malevolent fairy had set upon him in revenge and transformed him into a shriveled old man.

The psychiatrist who treated David wasn't certain of the diagnosis, but he knew that something was going on here besides catatonia, something even more malignant and threatening, something neurological. It reminded him of another disease he had encountered in the past. But what kind of illness would strike a young person so suddenly and, within the space of days, leave him bedridden, tremulous, cowering, and mute?

A neurological consultant agreed that David didn't suffer from schizophrenia. David's condition seemed to defy probability, as well as everything in the neurologist's previous experience. Add forty years to his age, and the diagnosis would not have been difficult at all: rigidity, tremor, weakness, slowness of movement, a masklike face, and in extreme cases mutism are the hallmarks of Parkinson's disease.

In 1817, James Parkinson, a general practitioner in Shoreditch, a pleasant resort town near London, wrote a short essay that described the "shaking palsy": "involuntary tremulous motion, with lessened muscular power, in parts not in action and even when supported . . . the senses and intellect being uninjured."

These three signs continue to be absolutely essential for the diagnosis. The patient displays a tremor while at rest; the movements are slow; and the muscles are rigid, resisting the examiner's attempt to move them. David had all these signs, and he had most of the additional signs that usually appear after the disease has progressed for several years. These include reduced facial expressiveness, the maintenance of fixed postures, and aphonia (reduced or absent speech).

The signs of Parkinson's disease were unmistakable, but David's course varied greatly from what is traditionally encountered with the illness. As Parkinson wrote: "So slight and nearly imperceptible are the inroads of this malady and so extremely slow in its progress, that it rarely happens that the patient can form any precise period of its commencement."

Parkinson encountered the disease frequently on the streets of Shoreditch, and he puzzled over its etiology; the lifestyles and histories of those afflicted with the disease were too varied to suggest any one cause. But never among his patients, whatever their histories, had there been a man of twenty-three.

David's illness was so similar to Parkinson's disease that, despite the unlikelihood of its developing so rapidly and in one so young, the consulting neurologist decided on a trial of L-dopa, a drug that is converted within the brain into dopamine, the neurotransmitter that is deficient in Parkinson's disease. The results were sudden and dramatic. Within days David was up and walking about. More important, he could speak, and the story he told had revolutionary implications for our understanding of the human brain.

Over the previous months David, a drug addict, had been working on developing a home-brewed heroin substitute. He wanted to produce the compound MPPP, which is chemically similar to meperidine, best known by its trade name Demerol. Although MPPP had shown great promise during the late 1940s as a major pain-killer, pharmaceutical houses had lost interest in it, perhaps because of side effects they encountered but had not reported.

Several underground laboratories took up where the legitimate laboratories left off. During the 1960s and 1970s, amateur and sometimes not-so-amateur chemists concocted MPPP in kitchens and basements and other underground "laboratories," in search of euphoria-producing agents.

There were several reasons why MPPP was of interest to these illicit manufacturers. The recipes for it involve unrestricted chemicals; the synthesis is relatively simple to carry out; and no laws were being broken since the drug had never been listed as a controlled substance—indeed, its existence had not even been officially recognized. The chemical skills required to make it were minimal; if you could follow a recipe for baking a cake, then you could manufacture MPPP. But there is one absolute requirement: the procedures must be followed *precisely* in regard to heating the mixture and the amount of acid added to it. Too much of either, and MPPP undergoes a subtle alteration in chemical structure and turns into a monster, the closely related but perniciously destructive MPTP.

After he had synthesized several batches of MPPP, David became increasingly careless. Most likely, like many addicts, he waited until a craving developed before he took steps, increasingly desperate and peremptory, to obtain his "fix." Whether he rushed the process or added too much acid will never be known. What is known is that after injecting some of the final batch, he developed the characteristic signs of Parkinsonism over the next several days.

In order to understand what happened to David, chemists at the National Institute of Mental Health (NIMH) repeated his protocol. It was clear that he had injected a mixture of MPPP and MPTP. A crude mixture combining these chemicals was injected into laboratory rats. The rats went into a short-lived catatonialike state, then reverted back to normal. They showed no permanent signs and no hint of Parkinsonism.

This posed a dilemma. The investigators were almost certain that David's use of MPTP was the likely cause of what had happened to him, but all their attempts to induce the illness in the laboratory failed. No one appears to have thought of the simplest explanation for this: that the rat might not be the right animal for the experiment. This insight did not come until three years after David's death.

David had remained Parkinsonian for over a year after his diagnosis, and he required steady doses of L-dopa in order to function. Apparently, he also continued his drug habits, because his death was cocaine-related. The autopsy revealed a loss of cells in the substantia nigra in his brain. His doctors could celebrate a grim victory: Despite the negative results with the rats, their claim that the toxic substance had worked its destruction in the same area as in those afflicted with naturally occurring Parkinson's disease proved correct. At least in this one tragic case, a link had been established involving MPTP, Parkinsonism, and the destruction of cells in the substantia nigra.

David's case demonstrated that the events occurring within the brain of Parkinson's disease sufferers are more complicated than had previously been realized. It showed that young people in the prime of their lives could suffer sufficient substantia nigra damage to produce a condition like Parkinson's disease. Further, a contaminated drug could generate that damage. The researchers at NIMH published their findings in the form of a single case study in the 1979 edition of *Psychiatric Research,* but they were unaware of other cases like David's and uncertain of what to do next.

Interest in MPTP was not limited to a drug-using college student from Bethesda. In 1982 "designer drug" chemists near San Jose, California, began supplying their customers with what they dubbed synthetic heroin. Apparently their technique was no better than David's, and MPTP turned up regularly in the drug samples confiscated by local authorities. These authorities were puzzled by the fact that the MPPP-MPTP mixture produced acute effects, including severe burning in the vein upon injection, blurred vision, tremors, a metallic taste, and jerking of the arms and legs accompanied by tightness and stiffness of the muscles. Why would anyone inject into himself more than once a drug that induced such unpleasant effects?

In order to answer that question, one must know something about both addiction and the illicit drug trade. As with any sales transaction, the drug purchaser must be on guard against flimflam and outright deceit. Many an addict has returned home with a bag of what he is certain is pure heroin, only to discover he's been fooled by a mixture of anything from sugar, baking powder, and precipitants of local anesthetics, to—when dealing with more un-scrupulous dealers—rat poison.

One way that dealers assure buyers that everything is "cool" is by adding a potent, short-acting substance that will convince an addict that he or she is receiving the unadulterated drug. The immediate effect of MPTP certainly accomplishes that: Within minutes after injection, buyers can be certain they aren't shooting baking powder or sugar. It is likely that the narcotic properties of MPPP masked some of the effects of MPTP, rendering the unpleasant reactions more bearable while still persuading users that they had injected an active drug. Finally, addicts had no rea-son to anticipate the most dangerous effect of all, drug-induced Parkinsonism, since it didn't develop for three or four days. An addict's temporal horizons extend no further than the time it takes to walk or run to the source of the next fix. The immediate mo-ment is all that matters.

Before long, the customers of drug suppliers in San Jose began arriving in local emergency rooms. One of the first, George Car-rillo, was taken to the Santa Clara Valley Medical Center, where he came under the scrutiny of a perceptive neurologist, William Langston. Impressed with similarities between Carrillo's illness and typical Parkinson's, Langston started treatment with L-dopa and issued a public announcement warning heroin addicts of the dangerous contaminant in their drug supplies. At the time, Lang-ston was not aware of the 1979 NIMH report of David's case, so he linked MPTP-induced symptoms with those of Parkinsonism independently.

"When I gave him the L-dopa it was like melting a block of ice," Langston recalls. Within days Carrillo had dramatically im-proved, as had several others who had come to the emergency room as the result of the public warning. But though these patients were soon walking and talking almost normally, they continued to require large doses of L-dopa to sustain their improvement. It was obvious that if they were to survive and have any chance at all of resuming a normal life, they would likely need daily administration.

The substantia nigra is named after its dark appearance, the result of pigmented cells containing neuromelatonin. What makes it so important is its role in the synthesis and transport of dopamine. As was noted earlier, in Parkinson's disease the number of dopamine-producing cells is dramatically reduced. (In fact, more than eighty percent of such cell bodies must be missing before the symptoms of the disease first appear.) Successful treatment of Parkinson's involves supplying the brain with the missing dopamine. Since it is dopamine-producing cells that are affected, not dopamine receptors, the deficiency can be made up by supplying pills containing L-dopa. Through a series of chemical steps, large neurons in the substantia nigra synthesize dopamine from L-dopa and its precursor amino acid, tyrosine.

A useful analogy is provided by Dr. Sanford P. Markey, of the section on analytical biochemistry of the National Institute of Mental Health. He likens the manufacturing and transportability of the substantia nigra to activities at a manufacturing and shipping terminus. After the raw materials are received, manufacturing, packaging, and shipping of goods to multiple purchasers takes place. Supply and demand will be related to how smoothly all these components of the process function. Problems in delivery may be due to a lack of raw materials, high rates of absenteeism in the labor force, an absence of demand, strikes, and so on. Likewise, a neuronal impairment may result from a deficiency of needed precursors (amino acids); a failure of enzymes to bring about a necessary conversion; or transportation problems from the cell body to the terminal. A host of different diseases can contribute to such failures, but in Parkinson's disease, Dr. Markey suggests, "the effects of the disease are like those of a war in which the manufacturing centers have been destroyed and part of their function is fulfilled by hardworking cottage industries." When the dopamine-producing cells of the substantia nigra are lost, a Parkinson's sufferer can neither synthesize nor transmit the needed dopamine.

Situations in which tragedy provides insight into brain function are not unusual. The greatest advances in our understanding of brain localization—the "geography" of such functions as seeing and hearing—have come from studies of the effects of wartime shrapnel injuries, and our knowledge of the motor functions of the spinal cord would still be incomplete had the polio epidemic never occurred. Were it not for the victims of synthetic heroin, Parkinson's would have remained of interest chiefly to the disease's victims, their families, and the neurologists specializing in the care of patients with degenerative brain diseases. Instead, thanks to the MPTP experience, brain scientists now have the means to learn more not only about the disease but also the role of the neurotransmitter dopamine and its receptors.

Assessment of dopamine functioning in the human brain is based on two indirect measurements. In the first, a sample of cerebrospinal fluid is analyzed and measurements are made of HVA, dopamine's breakdown product when it is metabolized. Patients with Parkinson's and MPTP patients both have depressed HVA levels relative to normal people. This indicates not only a deficiency of dopamine in these patients but a biochemical similarity between these two conditions.

The second method of assessment utilizes two sophisticated techniques: the radioactive tagging of chemical substances that can then be highlighted and seen using a technique called autoradiography; and PET (positron emission tomography) scanning, an elegant form of imaging that allows us to visualize structures and chemical activity in the living brain. When a radioactively labeled version of L-dopa is injected into a normal subject, those brain cells capable of storing dopamine take it up. The radioactively labeled L-dopa—the "raw material" in the shipping analogy—is delivered to the substantia nigra and converted to radioactively labeled dopamine; the dopamine is then passed along the shipping routes to the terminal storage site, the striatum. The striatum of a normal individual will "light up" in a bright red color in a PET scan image, thanks to the radioactively tagged dopamine stored there. In patients with Parkinsonism, whether natural or MPTP-induced, this does not happen. The disease has destroyed the substantia nigra cells, so they are not present to convert the L-dopa to dopamine. The absence of dopamine means, obviously, that none has been delivered to the striatum.

Radioactively tagged substances in neuroscience experiments are the correlates of double agents in espionage. Typically, a double agent gains entry to a target area of special interest by making himself or herself indistinguishable. Once access has been achieved, messages are relayed back, usually by means of some special code. By keeping abreast of the agent's location and monitoring progress as the agent moves from one location to the other, an intelligence agency can gain valuable information about places that are otherwise inaccessible. Just so, within minutes after giving labeled MPTP to experimental animals, the compound can be detected within their brains.

(Interestingly, while the brains of guinea pigs, monkeys, rats, and mice all concentrate the drug, the rate of its disappearance from the brain varies greatly according to species. At the end of twenty-four hours very little MPTP remains in the brain of the rat—the reason why scientists at NIMH failed in their attempt to duplicate David's illness in a laboratory rat. If they had chosen a monkey, they would have been successful, because 50 percent of an administered dose of MPTP remains in the brain of a monkey two weeks after administration.)

What is not yet well understood is the mechanism by which MPTP inflicts its damage. If neuroscientists discover this, they may eventually be able to stop the destruction of dopamine-producing neurons in both forms of Parkinsonism.

Once in the brain, MPTP is transformed into a new substance, MPP+. MPP+ is a toxic chemical that was developed as a herbicide, known as Cyperquat. It shares many of the toxic properties of paraquat, a plant toxin that was once suggested as a means to render marijuana less appealing. (If marijuana plants were sprayed with paraquat, it was reasoned, users would get violently sick, discard that particular supply, and renounce the use of marijuana altogether.)

Chemically, MPP+ differs from MPTP in one important respect: solubility. MPTP is fat-soluble and is therefore capable of moving in and out of a cell through the cell membrane, which itself is largely composed of lipid (fat). MPP+, in contrast, is water-soluble, and once it is formed within the nerve cell, it remains trapped there. The result is a high concentration of a known plant poison within brain cells for long periods of time. If MPP+ can kill plants, investigators asked themselves, what can it do to brain cells and how does it do it? In order to answer that question, they searched for the mechanism by which MPTP is bioconverted to MPP+. After traveling down several blind alleys, they discovered that the enzyme responsible for this conversion is monoamine oxidase type B (MAO-B).

Monoamine oxidase (MAO) is an old friend to scientists studying the chemistry of the brain. As we saw in Chapter 7, patients suffering from severe depression often improve after taking medication that inhibits the action of MAO in breaking down dopamine, norepinephrine, and serotonin into their metabolites. When this action is blocked, higher concentrations of these amines are available in the synaptic cleft for longer periods of time; hence the antidepressant effect. There are two types of MAO: A and B. It is type A that is linked to depression. Type B is associated with the MPTP-to-MPP+ conversion. If the action of type B is blocked, this conversion never takes place.

If an animal is given a MAO-B inhibitor *prior* to the administration of MPTP, none of the harmful effects occur. At the behavioral level the animal is protected from Parkinson-like symptoms, while at the microscopic level the usual MPTP-induced patterns of brain damage are not seen. Clearly it is MPP+ and not MPTP that is the villain.

But as in any good detective story, identifying the villain isn't everything. Hercule Poirot must do more than twirl his moustache and point out the murderer. He must also explain how the crime took place and lay out its chronology in narrative form. After neuroscientists identified MPP+ as the killer of substantia nigra cells, they set out to discover exactly how the lethal chemical did them in.

To pursue this they turned to paraquat, which, as we have noted, is structurally similar to MPP+. They discovered that this toxic substance acts by generating peroxide and other molecules that are able to wreak damage by attacking and destroying the cell membranes of substantia nigra cells. Why these cells principally? Although neuroscientists aren't certain, they suspect that neuromelatonin, the pigmented substance that is responsible for the darkened appearance of the substantia nigra, participates directly in MPP+ toxicity. It is speculated that neuromelatonin acts as a reservoir for these toxic molecules, which then set off a deadly cascade in which one chemical reaction stimulates another, compounding the destruction.

Whatever the mechanism, it is clear that MPTP, when transformed into MPP+, kills substantia nigra neurons in large numbers. Could drugs capable of protecting against MPP+-like agents prevent this destruction? Might this be a treatment approach for Parkinson's disease? The point is not that MPTP is the agent responsible for Parkinson's; no one claims that. But it is speculated that there may be a common mechanism at the molecular level that kills substantia nigra cells in both Parkinson's disease and MPTP poisoning.

Two treatment approaches have recently been developed to test this hypothesis. It was found that when Parkinson's disease patients are treated with the MAO-B inhibitor selegiline (Deprenyl), the progression of their illness is slowed. When L-dopa was given concurrently, in smaller doses than usual, its sometimes severe side effects were also reduced. This supports but does not prove the idea that MPTP-like toxins may be involved in naturally occurring Parkinson's disease. Presumably the effect works this way:

By inhibiting the effects of MAO-B, the conversion of a MPTP-like substance to its more lethal MPP+ counterpart is blocked. Unfortunately, although this theory is plausible, it is not the only possible explanation for what may be going on. It is equally plausible—and some neuroscientists would say *more* likely—that Deprenyl's positive effects stem from the reduction in the dose of L-dopa required to manage the illness.

These early efforts at treatment are somewhat crude and speculative in comparison with what we can probably expect over the next decade. Of equal importance are the changes in our thinking about some forms of brain damage that have been stimulated by the MPTP experience.

One question that arises is whether Parkinson's disease is a natural consequence of aging in susceptible people, and whether the rate of cell loss in the substantia nigra serves as the chief determinant of whether the illness will express itself. If it is, then drug users who have had low levels of exposure to MPTP—not enough to kill off sufficient substantia nigra cells to result in the Parkinson's syndrome—will likely develop Parkinson's disease as they enter their fifties, the time when some substantia nigra cell loss occurs in normal people. Presumably, if these addicts have fewer cells to start with as a consequence of their drug use, the loss of additional cells—a normal consequence of aging—will take a higher toll. Parkinson's disease will emerge in people who would otherwise have no particular susceptibility to the disorder. Studies carried out thus far by William Langston and associates support this premise.

But the most basic issue is whether the MPTP experience has broader implications. "The neurotoxicity of MPTP raises important questions about other chemical compounds," says NIMH's Sanford Markey. "Is MPTP a unique chemical, or could there be environmentally encountered natural products with similar neurotoxicity? Could related compounds be produced as the result of normal metabolism? Can MPTP-like activity be generated from otherwise harmless materials, either by activities unique to man like cooking, or from man-made substances like drugs or agricultural chemicals?"

Thousands of chemicals share the same structural elements of MPTP. By playing with its basic structure, adding and subtracting elements, and then testing for their killing effect on substantia nigra cells, neuroscientists have developed many analogues. Some of these are so toxic that the test animal dies before the MPTP effect has time to occur. It is hoped that ongoing research will shed additional light on Parkinson's disease. Until the MPTP experience, the disease generated little enthusiasm from most researchers.

It is ironic that from a contemporary social tragedy, drug addiction, would emerge important new insights into the workings of both the normal and the diseased brain. Who could have predicted that a brilliant and sadly isolated college student searching for new drugs and new highs would unintentionally provide neuroscientists with fresh ways of thinking about the life and death of nerve cells? Unexpected developments like this underscore the excitement of contemporary brain research; one can never be certain from which direction new knowledge will be forthcoming.

The MPTP story also underscores something else of great importance about chemicals and the brain: how narrow the margin is between benefit and damage. Only a slight modification of a chemical's structure, or how it is made, or its dosage, can unleash powerful unintended forces.

CHAPTER 10
GOOD DRUGS—
BAD DRUGS

*L*et me suggest for your consideration a naturally occurring liquid alkaloid, the color of pure water, that can be obtained anywhere without a prescription, yet is one of the most toxic substances known. It acts almost as quickly as cyanide; death ensues only a few minutes after swallowing a dose as small as sixty milligrams. Continued use of this legal substance in smaller, less toxic doses quickly leads to tolerance and dependency. According to a 1988 report, the pharmacological and behavioral processes determining addiction to it are "similar to those that determine addiction to drugs such as heroin and cocaine."

I am talking about nicotine, specifically nicotine in tobacco. Within the brain, nicotine mimics the neurotransmitter acetylcholine by acting at the acetylcholine receptor site and stimulating the nerve cell dendrite. Scientists took the first step toward learning this in 1889, when two investigators applied nicotine to a cluster of nerves in the neck of a rabbit. The nicotine prevented the transmission of impulses along the nerve fibers. It was later shown that because of its chemical structure, nicotine occupied a particular type of acetylcholine (cholinergic) receptor that scientists, not surprisingly, named the nicotinic receptor.

What is the origin of this association between the brain and nicotine? One possibility, admittedly speculative, is that nicotine evolved within plants as a method of protection against insect predators. When these predators ingest the nicotine in a plant, it blocks transmission along their acetylcholine-controlled nerve fibers and kills the insects. Nicotine is still used as an insecticide today in some parts of the world.

Much of our knowledge of nicotine and its receptors comes from investigations of the Amazon electric eel and electric fish, such as skates. The electric organs of these creatures contain a rich supply of nicotinic receptors. A ground-crawling denizen of the jungle, the deadly Malayan snake called the banded krait, also contributes to nicotinic receptor research because it manufactures a toxin-containing venom that binds to nicotinic receptors and blocks their action. The krait paralyzes its prey by attacking at the neuromuscular junction where the nerve stimulates muscles into action. All these creatures have provided neuroscientists with information about the structure and function of nicotinic receptors.

In order to study the nicotinic receptor in brain tissue, scientists at Georgetown University in Washington, D.C., synthesized acetylcholine and treated it with a radioactive label. When incubated with brain slices, this radioactively tagged acetylcholine attaches to the nicotinic receptors on the brain cells; these receptors are revealed by autoradiography, which images on special X-ray film the distribution of the radioactive acetylcholine bound to them. The more receptors present, the more exposed the X-ray film, and the more enhanced the autoradiogram. The process is a sophisticated version of putting a red hat on your toddler so that you can keep an eye on him and easily pick him out within a crowd of children in the playground. If you wished to keep track of more than one child, the technique would work just as well:

the red hats would cause all the children you're concerned about to stand out from the background.

In one early experiment, Kenneth Kellar and Rochelle Schwartz at Georgetown gave relatively low doses of radioactively labeled nicotine twice daily to rats and found that this treatment increased the number of nicotinic receptors in the brains of the rats. (A similar experiment in mice was carried out by Alan Collins and his colleagues at the University of Colorado, with the same results.) This finding came as a surprise to Kellar, since receptors usually increase in number when there is a *shortage* of stimulation and decrease when there is an excess of stimulation. A similar increase in the number of receptors was discovered in humans who smoke. A Scottish team compared the brains of deceased smokers with the brains of deceased nonsmokers and found that smokers had more nicotinic receptors than nonsmokers. In Sweden another study carried out in living subjects confirmed this.

Why would the regular use of nicotine increase rather than decrease the number of the brain's nicotinic receptors? Kellar and his team at Georgetown speculate that while nicotine exerts an initial and brief stimulatory or *agonist* effect lasting for seconds to minutes, it then desensitizes the receptors and thereby renders them nonfunctional for long periods of time, hours or even days. This stimulates cells in the brain to produce more receptors.

These actions follow what is known as a biphasic curve. Small doses of nicotine initially stimulate and mimic the action of acetylcholine, while larger doses and prolonged exposure result in a persistent depression of acetylcholine receptor activity. Smokers experience a biphasic response too. Low-level nicotine poisoning can be recalled by anyone who remembers smoking his first cigarette: nausea, dizziness, and general malaise. With additional smoking the user becomes tolerant to these effects and usually encounters the second phase: the pleasurable sensations induced by nicotine.

A study in cats showed that an intravenous dose rate that delivers the same amount of nicotine as a puff on a cigarette induces arousal in the animal's electroencephalogram (EEG), a recording of the electrical impulses in the brain. When the dose is changed—either lowered or increased—the arousal is replaced by a depression of EEG activity. This is an indication that there must be something special about that specific dose, since smokers could obviously puff harder or faster if they sensed a need to increase the amount of nicotine delivered to their brain. Cortical arousal and a general alerting effect may be the basis for nicotine's appeal. Studies involving human subjects show that nicotine increases mental efficiency, and the same effect is felt by nonsmokers who are given nicotine in the form of a tablet or chewing gum.

Once nicotine is in the brain, it leads to the release of a wide variety of neurotransmitter substances, including the endorphins, the body's own morphinelike chemicals implicated in pleasure and addiction, and dopamine, thought to be important to reward systems in the brain. (The endorphins are discussed more fully in Chapter 14.) Within the brain, certain cells involved in inhibition are nicotine-receptive, which may be the basis for nicotine's anxiety-reducing effects. Other effects include a lessening of irritation and aggression, and the suppression of appetite and loss of weight. With repeated smoking these effects diminish—the brain becomes *tolerant* to nicotine. In an effort to reexperience these effects in their original intensity, the smoker smokes more cigarettes, thus developing a physical dependence on nicotine.

The withdrawal symptoms that appear whenever the smoker attempts to cut down suggest that the brain has modified its functioning in response to regular nicotine "hits." Lacking nicotine, the brain is thrown off balance; it attempts to restore that balance by initiating craving on the part of the smoker. Indeed, craving is the behavioral and psychological correlate of events happening at the level of the nicotinic-acetylcholine receptor. In the absence of the stimulating effects of nicotine, the smoker who is attempting to stop can expect to encounter the following signs of nicotine withdrawal: craving, irritability, frustration, and anger; anxiety; difficulty in concentrating; restlessness; decreased heart rate; increased appetite and/or weight gain. According to the revised third edition of *Diagnostic and Statistical Manual* of the American Psychiatric Association (*DSM-III-R*), nicotine withdrawal is a form of organic mental disorder—in essence, a brain disease.

Is nicotine addiction curable? Studies carried out on people who have tried to stop smoking show that at the end of a year, somewhere between 70 and 80 percent have relapsed. No doubt one of the reasons for this alarmingly high relapse rate is the large amount of nicotine that the average smoker takes in. A pack-a-day smoker puffs at least fifty thousand times a year. That adds up to a lot of nicotine "hits" delivered to the brain, hits that cause the release of neurotransmitters that generate rewarding and pleasurable effects.

Nicotine's power to produce these positive mental and physical effects is the principal reason that its use isn't likely to disappear altogether—and perhaps shouldn't

Does this last statement seem insensitive to the fact that smoking is the primary preventable cause of death in developed countries today? That may be a fair objection, but there is some preliminary research that, if it bears fruit, may rehabilitate the status of nicotine and change it—at least in some respects—from a "bad" drug to a "good" one.

The fact that chronic use of nicotine stimulates the production of additional nicotinic receptors may have important implications for the treatment of Alzheimer's disease. This is because autoradiograms of Alzheimer's brains reveal a dramatically reduced number of cerebral nicotinic receptors. What would happen if nicotine were administered to patients suffering from Alzheimer's disease either by smoking or preferably in some other form such as a transdermal skin patch? Would they improve mentally as a consequence of more nicotinic receptors? So far, no one is certain. It is difficult to justify the ethics of giving patients substances that are known to be harmful. Even when given in a nonsmoking form, nicotine can cause adverse effects like irregularities in heart rate and high blood pressure. In addition, there are possible receptor problems at the molecular level. Kellar cautions that if a patient gets too much nicotine, it may initially stimulate these receptors but eventually lead to depression of their function.

One treatment approach might be to use an alternative nicotinelike substance that stimulates nicotinic receptors without desensitizing them as nicotine does. Or there might be a second drug, given concurrently with nicotine, that would block any antagonist effects. In this way nicotine could be useful to Alzheimer's patients without the risk of further decreasing the function of nicotinic receptors.

The point of this discussion is not to convince smokers to stop smoking or, worse, to warn them that they will be unable to do so should they try. Rather, I am out to change common misconceptions about "good" and "bad" drugs. It makes no sense to draw hard and fast "good and bad" distinctions between, say, nicotine and heroin or cocaine. According to the 1988 surgeon general's report, "The pharmacologic and behavioral processes that determine tobacco addiction are similar to those that determine addiction to drugs such as heroin and cocaine."

At the level of the neuron there are no good or bad drugs. There are only drugs that alter the brain, sometimes in ways that society chooses to label good, and at other times in ways that society thinks are bad. Social and economic factors also play a large part here. One of the reasons for cocaine's initial popularity and acceptance stemmed, I am convinced, from its ability to induce mental qualities that we highly value in this society: heightened energy, confidence, productivity, gregariousness, and seemingly insatiable enthusiasms for power, wealth, and sexuality. Only when the rueful discovery of cocaine's destructive effects on the personality became generally known did a consensus emerge that it caused more harm than good, and should therefore be designated a "bad" drug.

But to the psychopharmacologist cocaine, nicotine, heroin, and amphetamine are neither good nor bad. It's true that some of the immediate effects of the abuse of, say, heroin are more harmful than those of nicotine; a person's death from heroin often occurs within the first few months or years of use, rather than from lung cancer twenty years or more after taking up smoking.

But that is not sufficient reason to label one drug bad and another good. In England heroin is preferred over morphine as the drug of choice to alleviate the often unbearable pain of terminal cancer. Thanks to the judicious use of heroin, cancer victims there are able to live out their remaining time and die in dignity. Under such circumstances, can it be a "bad" drug? Isn't it a more sensible approach to consider *all* mind- and brain-altering drugs as good or bad only in the context of the use to which they are put?

One of the oldest and most widely used psychoactive drugs in the world provides another example of the importance of context. In 1674 a pamphlet appeared in London entitled "The Women's Petition Against Coffee, representing to public consideration the grand inconveniences accruing to their sex from the excessive use of the drying and enfeebling liquor." The authors complained that their men were drinking too much coffee and, as a result, were as "unfruitful as those Desarts whence that unhappy berry is said to be brought." They spelled out the effects on lovemaking of the "base, black, thick, nasty bitter stinking nauseous Puddle water."

This was quickly followed by "The Men's Answer to the Women's Petition Against Coffee," which asked, "Why must innocent coffee be the object of your spleen? Tis not this . . . that shortens Nature's standard, or makes us less Active in the Sports of Venus, and we wonder why you should take these Exceptions."

Obviously the men prevailed in this exchange; few people today believe that coffee has any appreciable effect on sexual desire or performance. But it does exert effects of no less importance. The ability of coffee and other caffeine-containing beverages to provide mental and physical stimulation is unquestioned today.

The caffeine in coffee, tea, and many soft drinks belongs to the chemical class of xanthines. *Xanthene* is a Greek word meaning "yellow" and refers to the residue that remains when xanthines are heated with acid and left to dry. Of the three xanthines, caffeine, theophylline, and theobromine, caffeine exerts the greatest effect on the brain.

In contrast to tobacco and the other mind- and brain-altering drugs mentioned in this book, it seems safe to assert that caffeine ingestion is almost universal. To avoid caffeine, a person would have to avoid coffee and tea and many medications, colas, and candies. Unless they have been decaffeinated, all these substances lead to activation of the cerebral cortex and an arousal pattern on the electroencephalogram. This results from the direct action of caffeine on the cortex and the release of norepinephrine in the brain. Not only is a person more alert while under the influence of caffeine but attentiveness is increased, along with an improved capacity to carry out repetitive, inherently boring tasks. Actual enhancement of performance occurs only in fatigued subjects, suggesting that caffeine is most helpful when a person is already tired; it exerts much less effect on rested, attentive subjects.

Social custom reflects neurophysiology; in general, the day's major portion of coffee is taken soon after arising. Later in the day, most people prefer caffeine-containing pick-ups such as sodas and teas. I regularly reach for a cola in order to keep myself writing, and I do find the caffeine helps, though not to the extent that it helped Balzac. "It causes an admirable fever," he wrote. "It enters the brain like a bacchante. Upon its attack, imagination runs wild, bares itself, twists like a pythoness, and in this paroxysm a poet enjoys the supreme possession of his faculties; but this is a drunkenness of thought as wine brings about a drunkenness of the body."

Knowledge of caffeine's mechanism of action dates back no more than a quarter of a century. This is remarkable, considering that caffeine-containing compounds have been in use since at least A.D. 900, when an Arabian medical book suggested that coffee could be used as an anti-aphrodisiac.

Within the brain caffeine, like some of the other xanthines to a lesser extent, blocks the brain's receptors for a substance called adenosine. Under ordinary conditions adenosine is thought to act as an inhibitory neurotransmitter. It is believed to produce sedation by inhibiting the release of other neurotransmitters of an excitatory type. Caffeine's stimulant action blocks the inhibiting action of adenosine, thus leading to stimulation, alertness, and an elevation of mood.

The imbiber of too much caffeine—generally ten or more cups of coffee a day, or their equivalent in tea or caffeine-containing sodas—starts showing signs of "coffee nerves," or caffeinism. The symptoms are strikingly similar to anxiety neurosis: a pervading feeling of restlessness, difficulty sitting still or composing one's thoughts, vertigo, headache, agitation, and an irregular heartbeat. Indeed, in some people a special form of anxiety, the panic attack (see Chapter 13), can be artificially induced by coffee.

A sudden reduction in habitual coffee consumption can also create problems, among them headache, irritability, restlessness, and drowsiness. Caffeine withdrawal headache is one of the most common causes of unexplained headache seen in neurologists' offices. Typically, the person complains of a dull or pounding headache that occurs eighteen to nineteen hours after the last "dose" of a caffeine-containing compound. Often patients come to the neurologist with suspicions that caffeine is the culprit, since they have observed that the headache can be relieved by taking in caffeine. Caffeine withdrawal headache is most likely due to the drug's ability to constrict blood vessels that feed the brain and scalp. When the caffeine is withdrawn, this vasoconstrictive effect disappears and is replaced by a rebound dilation of the vessels. It is this that causes the headache. (Dilated, pulsing blood vessels are also encountered in migraine, which often leads caffeine withdrawal headache to be misdiagnosed as migraine.) So common is this particular form of headache that many over-the-counter headache remedies contain caffeine.

As with all addicting drugs—and caffeine is addicting, even though we traditionally think of it as a "good" drug—tolerance is eventually achieved. After years of abuse, caffeine-containing substances may actually induce drowsiness rather than insomnia. A patient in my care several years ago drank one or two cups of strong coffee before going to bed in order to induce sleep!

Although it is less dangerous and comes from a different chemical family, caffeine's actions resemble those of amphetamine. In both cases, the user is "hyped up," fatigue disappears, mood improves, and sociability is enhanced. All of these effects result from caffeine's action on the brain, and it is very likely that new substances, based on the caffeine molecule, will be synthesized in the future that have similar desirable behavioral effects. Would such developments alter our tendency to put judgmental labels on commonly used substances that act on the brain? Such labeling is dubious, as we have seen in our scrutiny of the hazards of drugs culturally considered "good" or legitimized despite their adverse effects. It is time now to turn our attention more fully to some of the benefits that have come from "bad" drugs that our culture frowns upon.

CHAPTER 11
THE DUST OF ANGELS

In the 1950s anesthesiologists developed a new and promising anesthetic called phencyclidine. PCP, as the drug is now called, brought on a rapid loss of consciousness and reduced postoperative pain. In early tests with animals the drug seemed just about perfect; in contrast to most anesthetics, it did not suppress breathing or heart rate. In fact, it worked so well in monkeys—inducing what one experimenter described as a serene appearance—that in 1957 the manufacturer, Parke-Davis, began clinical trials of the drug as a human anesthetic under the trade name Sernyl. But within a few months, reports of bizarre incidents began filtering back to officials at Parke-Davis. Fifteen percent of patients who awakened from the anesthetic showed marked confusion, violence, and psychotic behavior indistinguishable from schizophrenia. By 1965, PCP-induced psychosis among patients participating in clinical trials reached such alarming proportions that Parke-Davis withdrew Sernyl as an investigational drug for humans. It remained on the market as an animal anesthetic and, when fired in syringe bullets from "tranquilizer" guns, as a means of immobilizing wild or dangerous animals.

At any other period this mind- and brain-altering drug would have reverted from people back to the animals on whom it had first been tested without incident, and the drug would have been heard from no more. But 1965 was no ordinary time; it coincided with wide enthusiasm in some quarters for psychedelic drugs. Moreover, PCP is easily synthesized—it can be prepared in a basement "laboratory" by anyone with access to easily available ingredients and chemical equipment, and an average understanding of high school chemistry. Before the year was out, the drug had appeared on San Francisco streets as the "peace pill," an eponym that deliberately invoked both mental tranquillity and Vietnam war politics. Other names followed, including hog, rocket fuel, crystal, and the Great Pretender (a reference to its frequent substitution for drugs sold as LSD, cocaine, and amphetamine). But the most popular name applied to it, then as now, was angel dust.

Coincident with the drug's street use, researchers in psychiatric hospitals were carrying out non–FDA approved experiments using PCP as a "facilitator" for psychotherapy. In line with the then-prevalent Freudian form of psychiatry, people who suffered from emotional disorders were thought to be "blocking," somehow holding back critical events in their pasts. PCP might help them, Freudian psychiatrists theorized, by releasing repressed or hidden thoughts, memories, and fantasies. Other psychiatrists welcomed the drug as a "model" for schizophrenia because it induced schizophrenialike effects. As hard as this may be to believe, they tested this theory by administering PCP to perfectly healthy volunteers, essentially to see what would happen.

By 1978, it had become obvious to the FDA that the drug was entirely too dangerous for human use and required restriction even with animals. This effectively brought to an end the drug's commercial value in legitimate channels. But street use not only continued but proliferated. In the early 1980s, PCP's effects on the brain were coming under close scrutiny in clinics and emergency rooms. Taken at low doses, the drug was found to induce a state resembling drunkenness, accompanied by numbness of the arms and legs. At larger doses the analgesic effect came into play: the person literally felt no pain. This had practical consequences, as the police across the country discovered. A person high on PCP wasn't deterred by pain-inducing restraints and experienced an unusual degree of strength, power, and endurance. Taken at even larger doses, the drug's psychic effects came to the fore. The patient—the proper term, since by this point medical attention had either been sought or, more often, forced upon the PCP user—acted and talked like a schizophrenic, experienced disturbing sensations similar to those described by individuals undergoing sensory isolation, and assumed and maintained bizarre bodily postures for hours at a time. On occasion, head-rolling movements, grimacing, and repetitive chanting speech were also observed. These effects lasted about four hours and were succeeded by depression, paranoia, and sometimes assaultive suicidal or homicidal behavior. Because of the absence of pain and the paranoid thinking, accidental injury was a frequent consequence of PCP use. Drownings were especially common, since PCP altered perceptions and produced sensory distortions. At even larger doses patients underwent epileptic convulsions and respiratory depression, ending in death.

Eventually, the dangers of PCP use became common knowledge on the street. But the appeal of a powerful mind- and brain-altering agent can never be underestimated. Users soon discovered that many of the serious side effects could be reduced, while leaving intact the mental rewards, if the drug was smoked rather than taken by mouth. In addition, PCP could be combined with marijuana for heightened effect, in a perilous combination known as "love boat." A decade ago I saw a patient who had killed his wife with a shovel after the two of them smoked marijuana that had been laced, unknown to them, with PCP. The last thing my patient could recall was seeing his wife assume before his eyes the form and identity of the Devil.

It would seem fair to say that by the late 1970s nothing positive had emerged as a result of the synthesis of this dangerous chemical. But insights into the human brain often come from unlikely sources, and in the case of PCP, the breakthrough occurred when neuroscientists started studying its effects on the brain. The first thing they learned was that PCP alters many neurotransmitters but does not appear to do so by acting *directly* on them. In 1981 neuroscientists discovered a specific PCP receptor. This did not provide an explanation of what PCP does within the brain, but it did provoke some daring speculations.

In fact, at one point scientists were seriously discussing the possibility that there was a natural PCP-like compound within the brain that might in some mysterious way malfunction and produce schizophrenia. Half-facetiously, they referred to this elusive compound as "angel dustin," but they were perfectly serious in hypothesizing that a natural substance or breakdown product within the brain might be responsible for schizophrenia. If a drug like PCP brings about such profound changes in normal brain functioning that a person under its influence may act for the first and only time in his life like a schizophrenic, they reasoned, PCP might plausibly be acting at the same site or sites within the brain responsible for schizophrenia.

But there was no way to confirm or disprove this perfectly reasonable hypothesis. Their research took an unexpected turn, however, when two naturally occurring amino acids that bring about excitation within the brain were studied. This research resulted in insights into how PCP could, under certain circumstances, be a "good" rather than a "bad" drug. In addition, it gave rise to knowledge that has revolutionized treatments for stroke and other destructive brain injuries.

+   +   +

Glutamate and aspartate are the two major excitatory amino acids within the brain. They are released from nerve terminals and act on receptors on the postsynaptic membrane. Currently it is thought that they depolarize the neuron—enhance its tendency to fire—by opening ion channels or, less frequently, by coupling via the glutamate receptor with other chemicals that initiate a chain reaction of interlinked chemical processes within the neuron.

Pathways that employ these excitatory amino acids are widely distributed throughout the brain; without glutamate and aspartate, brain functioning would simply cease. To this extent, if we thought of these substances as drugs, we would label them "good" drugs. But at the molecular level, "good" and "bad" are a matter of degree: Some very "bad" consequences can follow from too much or too little of a "good" drug.

In the late 1950s reports suggested that at certain concentrations glutamate is capable of destroying neurons. In 1957 two brain researchers reported that when given to mice, glutamate destroyed neurons in the retina of the eye and a portion of the hypothalamus within the mouse brain. Soon other reports confirmed that glutamate induced brain damage in a number of animal species, including primates. Moreover, analogues of glutamate proved neurotoxic in proportion to their excitatory action within the brain; that is, if a compound mimicked glutamate's excitatory action, it also possessed the ability to kill neurons if the dose were large enough. Something truly new had been discovered: At small doses amino acids like glutamate stimulate nerve cells to higher levels of activity, while at higher doses the same substances excite nerve cells to death. This in essence is the theory of excitotoxicity.

To discover why a natural brain substance is capable of destroying the brain's own cells, neuroscientists investigated the effects of certain analogues of glutamate, or compounds with similar chemical structures. These analogues included substances derived from seaweed, mushrooms, and the Quisqualus nut. But the most powerful analogue of all was a synthetic compound called N-methyl-D-aspartic acid, or NMDA, an excitatory neurotoxin that is even more potent than glutamate. When NMDA or any of the other analogues was placed into the brain, brain damage occurred in highly specific areas. As a rule, brain cells in nearby or even adjacent areas were spared. Analysis revealed that excitotoxins like NMDA exert their harmful actions in either of two ways: acutely, by interfering with the transport of sodium ions out of nerve cells, or slowly, by a lingering process that involves calcium transport.

It seemed logical to seek substances that would block the excitotoxic action of glutamate and NMDA and thereby prevent brain damage. So scientists turned their attention to developing antagonists to these excitotoxins. At this point, things took an unexpected and exciting turn involving the now all-too-familiar street nemesis, PCP.

When PCP is labeled with a radioactive tracer, it is seen to bind to a specific site on the glutamate receptor. Of even greater interest, it was discovered, PCP *protects* against damage brought about by the action of NMDA. Experiments in animals showed that if NMDA were given in the presence of PCP, the brain damage NMDA normally caused could be avoided. In other words, a dangerous synthetic drug that was used most commonly to get high possessed properties that protect the brain from excitotoxic damage.

Damage of this kind occurs after a stroke. One of the immediate consequences of a stroke is an outpouring of excitatory amino acids, especially glutamate, which induce brain damage beyond that resulting from the stroke itself. In the presence of PCP these harmful effects would probably not occur—indeed, PCP might well help limit the extent of brain damage after stroke.

Thus far, PCP's protective action against stroke has been demonstrated only in animals; for obvious reasons physicians treating stroke are reluctant to give a drug to human beings that will almost certainly induce hallucinations, agitation, and other dangerous and frightening effects. At this time, therefore, scientists are searching for a PCP analogue that will protect against glutamate-induced brain damage without causing these effects.

In an allied development, neuropsychiatrists are taking another look at PCP-induced psychosis. The discovery that PCP powerfully inhibits the receptor for NMDA strengthens a postulated connection between reduced functioning of the glutamate transmitter system and psychosis. Other theories suggest just the opposite—an increase rather than a decrease of excitatory neurotransmitters—as the cause of psychosis. Disagreements like this are to be expected, because normal functioning of the NMDA receptor depends on the interplay of many facilitative and inhibitory factors, and an imbalance of any one of them can alter the system's functioning.

Alterations in the amount of glutamate in specific parts of the brain may be responsible not only for schizophrenia and the damaging consequences of stroke but also for the brain injury secondary to low blood sugar, trauma, and seizures. An excitotoxic explanation for these and other conditions was suggested by several naturally occurring health disasters.

In December 1987 some 150 Canadians became seriously ill after they ate mussels that were later discovered to contain high amounts of domoic acid, a potent natural analogue of glutamate. Four people died, and among the survivors twelve suffered severe permanent memory loss similar to that of Alzheimer's disease. Autopsies on those who had died revealed damage to neurons in the hippocampus, one of the principal brain areas concerned with memory.

Based on the mussel poisoning–domoic acid connection, some neuroscientists are suggesting that Alzheimer's disease may be the result of excitotoxic assault. Supporting this theory is the finding that brain changes characteristic of Alzheimer's can be created in experimental animals. All that is needed is to inject an excitotoxin into the nucleus basalis of Meynert, where the brain's main source of acetylcholine-employing (cholinergic) neurons originate. The same result occurs if an excitotoxin is applied to those parts of the cerebral cortex to which the neurons from the nucleus basalis of Meynert project. In this case the excitotoxic effect works by injuring the cholinergic cells at their point of termination.

The excitotoxic theory is sufficiently paradoxical and bizarre—the brain literally destroying itself—to stir up a host of speculations. One favored by Washington University neurophysiologist John Olney is particularly troubling. In 1969 he showed that if glutamate in the form of its sodium salt (monosodium glutamate, or MSG) is given orally to rats or monkeys, it raises blood glutamate levels and damages the hypothalamus. MSG is, of course, not a rare chemical hidden away in a biochemist's laboratory; it is a substance that many of us who eat in Chinese restaurants ingest unless we specifically request that our food be prepared without it. It is an ingredient in many canned, processed, and frozen foods. (It must be listed as an ingredient on the package label, though other forms of glutamate also used as additives, such as hydrolized vegetable protein, may not be easy to recognize because the word *glutamate* does not appear in their names.) It would certainly never have occurred to most people to forgo MSG were it not for Olney's efforts over the past two decades to convince the Food and Drug Administration that the bad things brought about in animals by glutamate can happen in humans as well.

The claims that MSG produced excitotoxic injury in the developing brains of young children seemed sufficiently credible for the manufacturers to voluntarily remove the flavor enhancer from baby food in 1976. It still remains, however, in soups and other processed foods that are commonly fed to young children. Theoretically, the adult brain is less vulnerable to excitotoxic injury from food since for the most part its growth and development are complete. But even in adults, some parts of the brain remain vulnerable to all harmful chemicals that enter it. These are areas involved in memory and neuroendocrine activity, areas of concern in illnesses like Alzheimer's, Huntington's chorea, memory loss, and Parkinson's disease, to mention just a few conditions that may have an excitotoxic base.

Obviously not all individuals exposed to an excitotoxin go on to develop a brain disease. Neurobiology, like life in general, is rarely that simple. More likely, only some people in a highly heterogeneous population will prove susceptible to excitotoxic injury. Nonetheless, the ubiquitous use of MSG and its potential for harm are a continuous source of concern.

It is a strange and unlikely trail that PCP, so lethal itself, has revealed: the theory of neurotoxicity, new knowledge about the mechanism of neurotoxic injury after stroke and possible approaches to preventing it, and unanswered questions about the safety of a common food additive. The more we discover, the more we recognize that mind- and brain-altering drugs by their nature have boundless potential for good and harm.

CHAPTER 12
DR. FREUD AND THE DEVIL'S BARGAIN

Centuries ago, the Chinese discovered a new treatment for asthma: a tea made from herbs that they called ma huang. Asthma sufferers not only breathed more easily after drinking it, they felt refreshed and invigorated—though if too much were taken, they became tense and overstimulated.

The active ingredient in this Chinese folk remedy is ephedrine. It works by stimulating the sympathetic nervous system. This stimulation dilates the bronchial passages, which is good, but when taken in excess, ephedrine exerts sympathetic effects elsewhere in the body that are not. Patients given the drug often describe themselves as feeling "wired" and complain that they can't sit still, can't concentrate, can't stop their hands from shaking. All these effects are secondary to widespread stimulation of the sympathetic nervous system, which as we have noted controls heart rate, temperature, blood pressure, and respiration.

It was the Greek physician Galen who first detected nerve fibers that originated from what we now call the autonomic nervous system (described briefly in Chapter 2) and went to the stomach, heart, and other internal organs. Galen explained his observations using more poetry than science. He suggested these nerves carry the "sympathies," those visceral emotional reactions that are immortalized in such phrases as "his heart leaped with joy" and "she shed tears of sorrow" and "he ached with loneliness."

Eighteen hundred years later, the anatomy of the autonomic nervous system was finally worked out. It is now agreed that the system consists of fibers to and from the brain and spinal cord. The neurons that give rise to the sympathetic branch lie in the lower two-thirds of the spinal cord, in the thoracic and lumbar sections. The parasympathetic branch originates above this region of the spinal cord, in several brainstem nuclei, and below this region in the lowest, or sacral, region of the cord. As we noted previously, these two branches are not only anatomically separated but exert opposing effects on the body. Activating the sympathetic branch accelerates the heart rate, dilates the pupils of the eye, increases the breathing rate and sweating of the palms—in essence, creates an alerting response that prepares the person for flight or fight. The parasympathetic branch, in contrast, mediates the relaxation response: It slows breathing and heart rate, lowers blood pressure, and—as meditators can attest—induces feelings of relaxation and peace of mind.

Drugs that affect the sympathetic branch of the autonomic nervous system are useful in conditions marked by a decrease in sympathetic influence. Low blood pressure, for instance, can be treated by drugs that raise it by stimulating the sympathetic nervous system. But like the Chinese tea treatment for asthma, sympathetic stimulation is generated elsewhere in the body and can induce side effects serious enough to prevent a patient from using a sympathomimetic drug—one that stimulates the sympathetic nervous system. In the 1920s researchers came up with a drug that was very close structurally to ephedrine but that was chemically modified to avoid ephedrine's widespread sympathetic effects while still dilating bronchial passages in asthmatics. An added benefit of this drug—amphetamine—was that it could be inhaled, which meant it could be delivered directly to the lungs, the part of the body most affected by asthma.

Physicians who prescribed amphetamine noted over time that their patients experienced increased energy, a temporary elevation of mood, the need for less sleep and, curiously, a reduction in appetite. Soon these effects too were exploited, and amphetamine became the drug of choice for the treatment of narcolepsy, a bizarre neurological illness marked by the frequent and seemingly irresistible impulse to fall asleep, often at the most inconvenient and inappropriate occasions.

Narcolepsy had long been considered a neurological curiosity. Almost every medical school had its narcoleptic, who could be depended upon to fall asleep in full view of a roomful of fascinated medical students. But with the advent of amphetamine, the illness could be controlled, indeed almost eliminated, if enough of the drug were prescribed. Not surprisingly, heavy doses soon came into fashion, based on the "If a small dose works well, then a larger dose will work even better" school of therapeutics. In 1938, however, two narcoleptic patients treated with amphetamines developed an acute paranoid psychosis. Their symptoms so resembled an acute onset of paranoid schizophrenia that even skilled psychiatrists couldn't tell whether this was amphetamine psychosis or actual paranoid schizophrenia. The only way to make the differential diagnosis was to wait and see what happened to the patients. The cases of amphetamine psychosis ended after only a day or so, unlike those of paranoid schizophrenia, whose victims showed little or no improvement over the same time period.

At this point investigators stepped back and took a careful look at this new chemical. It was already in wide use. Over-the-counter amphetamine inhalers, called benzedrine inhalers, could be found in drugstores across the nation; people used them to relieve the stuffy noses and discomfort of colds. They also used them as pick-me-ups, stimulants, and mood elevators. By the early 1940s, benzedrine was being routinely dispensed to soldiers as a stimulant to keep them awake. According to a report in the *Air Surgeons Bulletin,* it "pepped up the subjects [fifty marines], improved their morale, reduced their sleepiness, and increased confidence in shooting ability." What could be the chemical explanation for such profound behavioral and psychological changes?

Although chemical formulas are kept to a minimum in this book, amphetamine's structure is sufficiently intriguing and enlightening to merit an exception.

It is clear from the structural formulas for amphetamine, dopamine, norepinephrine, methamphetamine ("speed"), and ephedrine that the general chemical structure of all these compounds is almost identical.

The similarity between amphetamine and two of the brain's most important neurotransmitters, the catecholamines dopamine and norepinephrine, suggests that amphetamine works by mimicking their action. This can also be said of ephedrine, since the structures of amphetamine and ephedrine are almost identical. Comparing the formulas of the old Chinese remedy for asthma, ephedrine, and amphetamine reveals only one difference: the second carbon off the catechol nucleus carries a hydroxyl group (HO) rather than a hydrogen atom. This seemingly slight variation is responsible for an important difference. Ephedrine cannot easily cross what is called the blood-brain barrier and enter the brain; therefore it remains largely outside it, producing mainly peripheral effects such as dilating the bronchial passages.

Amphetamine, in contrast, easily enters the brain, where it works in areas responsible for mood, alertness, and energy levels. Amphetamine can be even further "hyped up" by chemical modification that changes it into methamphetamine—speed. Notice that methamphetamine possesses a CH_3, or methyl group rather than an H group on its tail end. This methyl group facilitates passage into the brain and thereby enhances the drug's potency.

Based on these observations of amphetamine's molecular structure, a theory of its actions emerged. Since amphetamine mimics the structure of norepinephrine, it seemed reasonable to suggest

its alerting effects probably arise from stimulating the norepinephrine neurons. The central site of these neurons is the locus
ceruleus, a tiny nucleus in the brain stem whose nerve cells project
upward.

This central coordinating center sends more fibers to more
parts of the brain than any other single nucleus. The fact that the
predominant terminal points of these nerve fibers are the cerebral

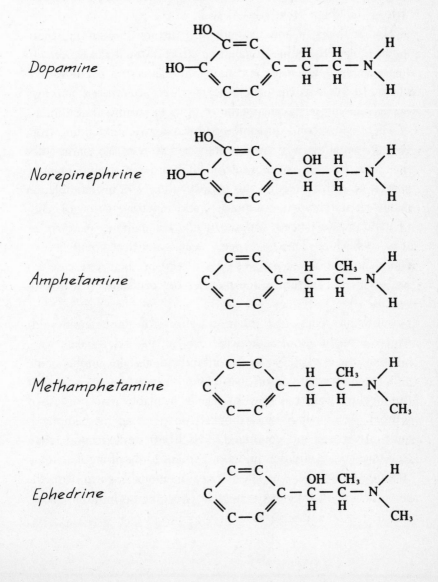

cortex and the hypothalamus lends additional support for the belief that this bundle of nerve fibers is responsible for amphetamine's alerting and antifatigue response.

Fibers of dopamine-producing neurons ascend from another
brainstem nucleus, the substantia nigra (which is involved, as we
have seen, in Parkinson's disease) up to various centers of the
cortex and limbic system. Stimulation of this mesocortical limbic
system produces euphoria and increased mental and physical activity—the "high" that abusers seek.

Higher doses of amphetamine and further mesolimbic stimulation replace the enhanced alerting effect of the lower doses with
hypervigilance, agitation, and the components of a paranoid psychosis. In this state the patient hears voices, sees visions, and may
experience himself as the victim of plots to harm or kill him.

There is a second dopamine pathway as well, extending from
the substantia nigra to parts of the brain responsible for smooth-
ness and continuity of movement. Higher doses of amphetamine
exert a powerful effect on this pathway too. Here overstimulation
results in stereotyped, compulsive, and repetitive behavior. Amphetamine's enhanced variant, methamphetamine, exacerbates
these behaviors. Thus, a "speed freak" may feel compelled to
count the number of grooves on a record or, as in one case described in the literature, the number of cornflakes in a box of
cereal.

Public awareness of the harmful effects of amphetamine ultimately brought about a dramatic drop in its use. Inhalers were
taken off the market. No longer did individuals take amphetamine
pills to combat fatigue or give themselves an energy boost. But
this decline was true only of legally available and prescribed
sources. Among individuals hooked on it or on methamphetamine, street demand continued. This heavy underground usage
continues today, and new and more potent forms of amphetamine
continue to emerge. The latest, "ice," a smokable form of methamphetamine, competes for popularity on the underground mar-

ket with intravenous forms of the drug. For users, the method of administration is of less importance than the endpoint.

Attempts to stop taking amphetamines result in a "crash" characterized by lethargy, fatigue, nightmares, headaches, a loss of self-control, outbursts of rage, and even physical violence. The most common symptom is depression, which peaks between forty-eight and seventy-two hours after the last dose and may last for several weeks, sometimes reaching suicidal proportions. Since a dose of amphetamine reverses this depression and produces a new and more intense high, the compulsion to take the drug once again is extremely strong. Eventually the cycle of euphoria/depression/euphoria dominates the user's life.

This behavioral-level cycle corresponds to events at the molecular-chemical level of the brain. Because of its structural resemblance to norepinephrine and dopamine, amphetamine can cause the release of these two chemicals from their presynaptic nerve terminals, thus increasing their concentration in the synaptic cleft. This release is further compounded by the fact that amphetamine interferes with the reuptake processes for these neurotransmitters. By enhancing the release and retention in the synapse of dopamine and norepinephrine, particularly within neurons involved in the pleasure response, amphetamine makes the user feel energized and "up." Eventually, when the supplies of these pleasure-mediating neurotransmitters are depleted, leaving neurons deprived of them, the addict enters into a deep depression.

Amphetamine is uniquely valuable to neuroscientists because it provides them with something for which they have long searched: a mind-altering drug whose mechanism of action in the brain can be explained. The window it offers on normal brain action suggests that when we feel elated or in high spirits (in "natural highs"), the catecholamine system is involved, while decreases in catecholamines can be associated with depression.

✣ ✣ ✣

Vying with amphetamine as the chief mind-altering agent of the 1970s and 1980s was a drug that was in use hundreds of years before amphetamine was developed. Cocaine is an alkaloid derived from the shrub *Erythroxylon coca*, found in the eastern highlands of the Andes Mountains. For over fifteen hundred years the indigenous populations have chewed coca leaves to enhance their strength and endurance. They even placed coca leaves within graves as a necessity for the afterlife. Cocaine was first extracted from the coca leaf around 1855. This availability of a supply of pure cocaine, along with the invention of the hypodermic syringe at about the same time, meant that the drug could be delivered in higher or greater doses than had ever been previously considered possible. Almost immediately, scientists sought new and different uses for it beyond enabling people to work harder and longer. Two physicians, later recognized as giants in their respective fields, surgery and psychiatry, spearheaded the effort to discover new therapeutic uses for cocaine.

Dr. W. S. Halsted, the father of modern surgery, found that injecting cocaine into an area near a nerve could block transmission within nerve fibers. This led to the use of cocaine as a topical and local anesthetic—the only purpose for which the drug is still used in medical practice today. It works by altering the nerve cell membrane. It is believed that cocaine molecules dissolve in the lipid matrix of the membrane, where they bind to receptor sites within the sodium channel. This interferes with the opening of the channels and the transfer of sodium across the membrane. Since sodium cannot move across the membrane, the axon cannot be depolarized, thereby blocking the nerve impulse.

At about the same time, the young Sigmund Freud was experimenting with cocaine as a stimulant, an aphrodisiac, and a cure for depression, alcoholism, asthma, and morphine addiction. This last use almost ruined his reputation and career before they were launched. He referred to cocaine in a letter to his fiancée as the "magical drug," and addressing a group of psychiatrists in 1885, he confidently spoke of using cocaine to alleviate "the serious withdrawal symptoms observed in subjects who are abstaining from morphine and of [its] suppressing their craving for morphine." But a short time after he made this statement, Freud was forced to retreat from it. He administered cocaine to a friend who was suffering from morphine addiction, and after spending a fearful night nursing his patient through a severe bout of cocaine-induced psychosis, Freud changed his views and stopped issuing extravagant claims about cocaine as a cure-all.

Doctors and scientists were not the only people fascinated with the potential uses of cocaine. Entrepreneurs, encouraged by the sanguine claims of Freud and some of America's leading medical authorities about the safety of the drug, vied with one another to produce cocaine elixirs for the masses. Perhaps the most famous purveyor was Angelo Mariani, who manufactured Vin Mariani, an expensive mixture of wine and coca. Regular consumers of this concoction included Thomas Edison, Sarah Bernhardt, and the Prince of Wales. Science-fiction writer Jules Verne enthused, "Since a single bottle of Mariani's extraordinary wine guarantees a lifetime of a hundred years, I shall be obliged to live until the year 2700."

For those given to less exotic associations—and with budgetary constraints—there was Coca-Cola. In the early years the soft drink was indeed the "real thing"; it contained enough cocaine to induce a pleasant lift if not always an actual "high." Although cocaine was removed from Coca-Cola in 1903, it is still flavored with decocainized coca leaves. (The Coca-Cola Company continues to purchase from the government the residues resulting from the preparation of all legally produced cocaine compounds in this country.)

By the late nineteenth century, neighborhood soda fountains were dispensing cocaine-containing drinks, among them such favorites as Wise Ola and Koka Nola, not to mention a perennial favorite with the simple name of Dope. At the height of the cocaine craze the drug was included in many patent remedies and tonics, often without any labeling that would alert the user to its presence.

This changed in 1906, with the passage of the Pure Food and Drug Act. Cocaine's popularity had been accompanied by increasing reports of the drug's ability to cause dependence. In 1896 an investigation by the Connecticut Medical Society concluded that cocaine treatments for hay fever and other minor illnesses had created major drug dependencies in many patients. It ended with the statement: "the danger of addiction outweighs the little efficacy attributed to it." On the literary front, the debilitating effects of cocaine were depicted by Sir Arthur Conan Doyle in *The Sign of the Four*. After Sherlock Holmes injects himself with a cocaine solution in order to achieve "mental exaltation," Watson asks a series of questions that touched on issues of increasing public concern at the time.

> Count the cost! Your brain may, as you say, be roused and excited, but it is a pathological and morbid process which involves increased tissue-change and may at least leave a permanent weakness. You know, too, what a black reaction comes upon you. Surely the game is hardly worth the candle. Why would you, for a mere passing pleasure, risk the loss of those great powers with which you have been endowed?

In 1914 reports of drug dependencies and deaths led to the inclusion of cocaine among the drugs prohibited by the Harrison Narcotics Act. This federal act barred the use of cocaine in patent medicines and made recreational use of the drug illegal. Although this was all to the good, the government's action was a classical example of being right for the wrong reasons—it incorrectly classified cocaine as a narcotic rather than a stimulant.

The act effectively drove cocaine underground, but it didn't disappear by any means. A small but influential segment of the population, particularly bored and wealthy young people, continued to use it into the 1920s. In the 1930s amphetamine replaced cocaine as the drug of choice; but this sequence reversed itself in the 1970s and 1980s, when cocaine returned as the "champagne of drugs." Indeed, cocaine and amphetamine have regularly vied for first place in the category of stimulant drugs of abuse. In order to understand why, it is necessary to examine the nature of addiction and the events that occur within the brain that deprive addicts of their freedom.

At the basis of addiction are certain neurotransmitters, most notably dopamine. Within the midbrain, well below the cerebral cortex, is a cluster of dopamine-containing cells called the ventral tegmentum. For years, neuroscientists have known of the importance of the ventral tegmentum in mediating maternal behavior. Injuries in this area will interrupt normal caring actions on the part of a mother rat toward her young. The area is also known to be an important mediating area for addiction. Laboratory animals can become addicted to drugs, and when specific injuries are induced in test animals in some of the dopamine projections from the ventral tegmentum to the higher centers in the limbic and frontal brain regions, the animal is more likely to self-administer amphetamines and other drugs of abuse. (In the experimental setting, there is a button connected to a machine that delivers amphetamine, and the animal will push the button repeatedly in order to continue receiving the drug.) It is conceivable that some as-yet-unspecified injury or abnormality within the dopamine system could make a person more vulnerable to the reinforcing actions of certain psychoactive and addicting drugs.

As studies have revealed, the ventral tegmentum plays an important role in cocaine addiction. Dopamine neurons in the ventral tegmental area are connected by a bridge of fibers to another brain structure called the nucleus accumbens, which mediates the pleasure response. The presence of cocaine inhibits dopamine reuptake, and more dopamine remains in the synaptic cleft. The excess overstimulates the nucleus accumbens, producing a pleasurable reaction. The user may sustain it by taking more cocaine. Virtually all drugs abused by humans—including cocaine, opiates, amphetamine, alcohol, nicotine, and marijuana—lead to increased levels of dopamine in the nucleus accumbens. To this extent, all these drugs share a common mechanism of action in inducing a "high." But cocaine communicates *directly* with the brain's pleasure centers, and this prolonged action, the source of the user's "high," helps explain why it is perhaps the most addicting substance in the history of the world.

This stimulation provides a cocaine user with a nonstop pleasure experience. Everything else seems dull and lifeless. Rather than extracting pleasure from everyday experiences, the user opts for what one researcher has called "brain hedonism." For the addict who is trying to come off cocaine, the state of anhedonia—the loss, seemingly permanent, of all capacity for normal pleasure—seems most frightening of all.

Cocaine is taken in several ways. With intranasal administration, the onset of its action is within two minutes. Intravenous use and freebase smoking both produce almost immediate effects. In general, the faster the cocaine reaches the brain, the more addictive it becomes. In fact, when drugs are ranked according to their intrinsic (primary) reinforcing powers—their direct pharmacological effect on a central neuronal circuit that mediates pleasure—cocaine and amphetamine come out on top. This may come as a surprise—most people believe that heroin and other narcotics are the most addictive drugs. But animal research, as well as experience with the most sophisticated animal of all, the human being, indicates otherwise. Provided that the route of administration is the same for all the drugs being assessed (especially with intravenous use), cocaine and amphetamine are the two most addictive drugs.

Over the last several years, investigators have learned a lot about how cocaine works, particularly how it produces its high. Smoking a drug like crack in a pipe is one of the most rapid and efficient drug delivery systems known. Smoked cocaine reaches the brain so quickly that for all practical purposes it is no different from the speed of delivery from a needle. How fast is fast? Probably in the neighborhood of three to five seconds. This speed has important consequences, since the subjective effects of cocaine are related not only to the blood level achieved but to how quickly that blood level is reached. When inhaled, crack cocaine goes directly to the lungs, where it is readily taken up by the blood, pumped once through the heart, and express-delivered to the brain.

Cocaine alters the autonomic nervous system by increasing the breathing rate, the heart rate, and blood pressure (the reason cocaine may cause a stroke or heart attack in certain users). Hunger, depression, and fatigue melt away. The user feels vigorous, hypersexual, friendly, and invulnerable. Many comedians and entertainers have employed these cocaine effects to fuel their creativity. Word associations, puns, quips, the instantaneous creation of inspired sketches—all of these can be enhanced under the influence of cocaine. But unfortunately, as prominent show business performers and others have found out, there is a darker side to cocaine—the "crash" that follows the euphoria and the conviction that one is above and beyond all of life's frustrations and disappointments.

Still, not everyone who takes cocaine or other addictive drugs goes on to become an addict. Why can some people take the drug and leave it, while others can only take it and take it and take it again? There are no absolute explanations for the extremely important question of why this is so, but recent findings and intriguing hypotheses are illuminating.

Researchers have turned up abnormalities in the brains of those who suffer from addictions. These abnormalities involve imbalances in brain chemistry.

Imagine a man who from his earliest years suffers from a low mood. Sometimes he's just "blue," but on other occasions he's definitely depressed. Now, imagine that that man tries cocaine or amphetamine. Either drug increases the level of available dopamine. Accompanying the rise in dopamine is an elevation in mood. What might one conclude about the chemical nature of this person's depression and his risk of becoming addicted to cocaine or amphetamine?

Actually, this hypothetical example is typical of patients studied at Harvard Medical School. According to Dr. Edward Khantzian, a psychiatrist at Harvard, some cases of depression *are* the result of low levels of dopamine in the brain. This can be corrected, and often is in treatment, by antidepressants that raise the level of dopamine. But our hypothetical patient doesn't know anything about that. He doesn't realize that he suffers from an illness and would never think of seeing a doctor about his feelings. All he knows is that a dose of cocaine or amphetamine makes him feel the way he would like to feel: much less "down," much more satisfied with his life. This man's risk of addiction to cocaine is extremely high since, in a real sense, the drug is bringing about a "normalization" of his brain chemistry along with his mood.

Picture a hard-driving workaholic public relations executive. He's charming and ebullient, never out of sorts, always "up." Such a person would seem to be at low risk for developing a cocaine habit, especially if his dopamine level is *higher* than normal. But according to Dr. Roy King, a psychiatrist at Stanford Medical School, "If you're highly extroverted, even slightly manic, by temperament, cocaine augments your natural high." Dr. King said in *The New York Times,*

> For people who are naturally bubbling with excitement, crack seems to intensify the normal biology of a state they seek: it's an exhilarating, exciting high. Cocaine makes more dopamine available to the brain, and they already have higher levels of dopamine. Smaller amounts of cocaine would have greater effects with them, making it more rewarding than with most people. They become more addicted to it because they get more intense pleasure from a given dose. That makes crack, the most intense form, especially appealing.

Addicts hooked on heroin or other narcotics may have an *under-*supply of a brain chemical. If the body's own natural opiates, the endorphins (about which we will have more to say in Chapter 14), are in short supply, the person may feel perpetually pent-up, filled with a smoldering rage. When such a person takes heroin, the increase in available opiates makes him feel calm, self-possessed, less aggressive. In a case like this, the opiates may function to dampen the brain circuits involved in aggression.

One of the most intriguing aspects of addiction is the mechanism whereby considerations of health, safety, and even survival become less important to the user than obtaining supplies of the drug. One thing seems certain: Drug seeking and drug using, as behavioral expressions, must originate within the brain. In order to have an effect, an addicting drug must act on the cells and molecules of the brain.

But the mechanisms responsible for the actions of addictive substances are incompletely understood. Complicating any determination of this important issue is the sticky matter of definition.

For years experts made a distinction between physical and psychological dependence. According to this theory, which is no longer accepted, intense physical dependence in an addicted person differs from psychological dependence. But this distinction often proved useless when applied to the real world. For several months a cocaine addict insisted to me that she could stop if she wanted to. Typically, she would stop using cocaine for a day or two and then, in an orgy of self-congratulation, double her use over the next several days. She had trouble sleeping, ate little, and experienced a chronic restlessness—all signs of physical dependence. Nor could she concentrate or stop herself from thinking or talking about cocaine. Eventually, she exhibited three components of drug dependence: an inability to stop herself from using it; continued use despite harmful consequences; and tolerance, the need for increasing amounts of the drug to produce the same behavioral effects. Was this woman, who was only in her mid-thirties, physically or psychologically dependent? Obviously, she was both. Physical and psychological dependence are intertwined.

Another mistaken notion that one hears frequently even today is that some kind of hidden psychological vulnerability is required in order for a person to develop an addiction. Experiments in animals suggest otherwise. If potentially addictive drugs are given to animals chosen at random from a colony whose members have no previous drug experience, these animals will quickly learn to push a button to obtain the addicting substance. Soon the pattern of drug use will be similar to that observed in humans. It is humbling as well as terrifying to realize that under the appropriate circumstances—the forced administration of an addicting substance within the setting of an uncontrolled and diabolical experiment—we could all be turned into addicts.

But is there, then, no such thing as an addictive personality? Are there no individuals who, if they do not get hooked on one drug, will end up abusing another? If there are, the likelihood increases that addiction has a genetic basis. It is possible that biological patterns like the low dopamine levels within the brains of certain users of stimulant drugs may be the result of a genetic susceptibility.

Two decades ago, such a hypothesis would have been the equivalent of saying, "It's all hereditary and therefore nobody can do anything about it." But corrective measures for genetic defects are no longer the stuff of science fiction. Today researchers and doctors are confident of the imminent prospect of altering gene structure in order to cure illnesses and diseases. If this can be done for appropriate individuals suffering from addiction, an important aspect of the mind will be transformed by indelibly altering the chemistry of the brain.

What certainly does appear to be true is that the *strength* of an addiction differs markedly from one person to another. It is estimated that only one cocaine user out of every four or five will become a chronic user or addict; with crack cocaine, the figures may rise to one person in three. An important variable is how easy or difficult it is for the user to get the drug. When access to drugs is limited, the likelihood of developing dependence is decreased:

obviously, you can't become addicted to or escalate the dose of something you can't get.

The process of addiction is not at all straightforward. People take drugs for different purposes and under different circumstances. For example, addiction to benzodiazepines, Valium-like drugs, may depend on negative reinforcement—that is, individuals may take the drug on a regular basis in order to feel less anxious, to get rid of bad feelings, not for any high or euphoria. Nicotine, too, is used for negative reinforcement; the smoker lights up in order to avoid unpleasant withdrawal reactions, not to get high. By contrast, a drug like cocaine is likely to be taken for the highly pleasurable feelings that it engenders, a positive reinforcement.

But the mechanisms that motivate users may change over time. Originally the person may start taking cocaine for the good feelings it produces. Later the reason may change, as the user reaches for cocaine to stave off the negative feelings engendered by the withdrawal experience. Indeed, the distinction between negative and positive reinforcers underlies one of the still-unresolved questions about why people take mind-altering drugs: Do they do so to feel good, or to prevent unpleasant feelings?

At this point, it appears that both factors play important roles at different times in different people, and even in the same person at different times. The circuits involved in mediating pleasure are not the same as those that fend off painful feelings, but in all cases, cues from the internal and external environment are important in linking the positive and negative feelings associated with drugs. This is a learned response. Neuroscientists know this because drug use and abuse can be dramatically altered by changes in the external environment, such as the presence of other drug users and drug paraphernalia, or frequent discussions of drugs. Internal states act similarly; the drug user learns to respond to depression or other bad feelings by reaching for a drug in order to feel good again.

At present, neuroscientists are operating on the assumption that for each addiction there are specific neurotransmitter irregularities. Correction of these irregularities may on occasion involve the use of antidepressants; the most popular drug currently used to combat cocaine addiction is an antidepressant. At other times, medicines may be employed that are addictive themselves but replace a more serious addiction. Ritalin (methylphenidate), which has been used for many years to treat hyperactivity, is an example. Ritalin increases the level of dopamine at the synapse. Although abuse of this drug is no trivial matter, most researchers would agree that given the grim choice, a dependence on Ritalin is unquestionably preferable to a cocaine habit. Ritalin abusers don't experience as difficult a withdrawal as those hooked on cocaine; it is a legal drug available through normal medical channels; its chemical purity is assured; and even regular use rarely leads to intravenous administration.

Advocates of addictive replacements like Ritalin believe that many if not all addictions result from people's efforts at self-medication. Those who have low levels of dopamine seek out and employ substances that raise the levels of dopamine in their brain. It seems most likely in the near future that additional treatments will be developed that employ medicines like Ritalin, which although potentially addictive don't pose nearly as many problems for the user or society as the drug of abuse. Thus far, no one has come up with any substance that replaces an addictive drug which is not addictive itself.

✛   ✛   ✛

Paralleling the desire for mental excitation, the achievement of "highs," is the desire for mental tranquillity. On many occasions we want not to be hyped up but relaxed, "mellowed out," or sedated. The decrease in physical activity that comes with relaxation is accompanied by a reduction in brain activity. Instead of large portions of the brain firing off nerve impulses, there is less neuronal activity, involving fewer neurons.

Modern attempts at inducing sedation-relaxation started with chloral hydrate, a drug first employed in 1869 and, to a limited extent, still in use today. Addiction to chloral hydrate was recognized by Dr. Benjamin Richardson, the person who introduced the drug into medical practice. Chloral hydrate addicts, Richardson wrote, were no better than "alcohol intemperants and opium eaters." The most infamous use of the chemical involved its administration to unsuspecting sailors in the form of a Mickey Finn. When chloral hydrate is added to an alcoholic drink, the imbiber soon lapses into unconsciousness. Many a sailor awakened from his chloral hydrate–induced stupor on a ship headed for the Orient.

Paraldehyde, another sedative, was first synthesized in 1829 but not used until 1882. Like chloral hydrate, paraldehyde still finds occasional use today, in the treatment of epileptic seizures. Its antiepileptic effect is based on the drug's ability to exert an inhibitory action on neurons and thus prevent seizure discharges from spreading from one area of the brain to another. Addiction, when it occurs, is easily recognized since the drug imposes a sweetish, cloying odor on the breath.

Less treacherous methods for inducing sedation and sleep date from 1903 and the introduction of the barbiturate barbital. Since then, more than 2,500 barbiturates have been synthesized. They differ from one another in speed of onset and duration of action. In general, those that are soluble in fat have faster onsets of action because they more easily enter into and pass through the lipid-rich membranes of the brain; their effects wear off more quickly than those less soluble in fat. Whether such a drug functions as a sedative or a hypnotic (sleep-inducing agent) depends upon the dose. A dose of thirty to fifty milligrams of phenobarbital taken during the day takes the edge off anxiety-provoking events. At double or triple the dosage, that same barbiturate will perform later in the evening as a sleep inducer.

Although still used as sedatives and sleeping pills, barbiturates are chiefly used now as anticonvulsants because they "dampen" the excitability and spread of nerve impulses and thus lessen the

likelihood of an epileptic seizure. But this use too is increasingly
rare, since more specific and more powerful inhibitors of neuronal
discharges have been developed that can stop seizures without the
dangers of dependence, addiction, or overdose. The combination
of barbiturates with alcohol is a particularly lethal mix, one of the
reasons these drugs have fallen out of favor as sedatives.

From the turn of the century until the 1940s, doctors and patients yearned for a drug capable of inducing relaxation without
decreasing alertness or lowering mood. To those living in the Age
of Anxiety, the prospect of an effective and improved antianxiety
agent exerted tremendous appeal. It still does. The mind-altering
agents discussed so far may interest only some portion of the population, but an antianxiety agent holds universal appeal. Is there
anyone who is not anxious, sometimes on a fairly regular basis?
Indeed, given the conditions of modern life and a world that is
becoming increasingly complex, demanding, and uncertain, anxiety seems, as some claim, a perfectly normal response. If this is
so, the search for a safe and effective antianxiety agent takes on
a particular importance.

CHAPTER 13
MOTHER'S LITTLE HELPER

*A*nxiety probably evolved in tandem with the evolution of the human brain. The capacity for imaginatively creating and responding to dangers that never materialize is unique to our species. In contrast to animals in the wild, who must manage real threats to their physical integrity, the threats that most of us encounter most of the time are of a subtler psychological sort. Anticipation of a loss of face in front of others, of one's job or financial status being at risk, of making a bad impression, of a whole variety of problems—these provoke anxiety.

Over the centuries people have thought about anxiety in many different ways. The English philosopher John Locke considered it a prime motivator for human behavior, a view that lives on today. Teachers, administrators, and supervisors are often heard to say such things as, "People have to be kept on their toes. Let them relax too much, and they become unproductive. A little anxiety is good because it keeps everyone motivated." But even advocates of this approach agree that when anxiety continues, the supposed benefits wear thin; people can no longer work efficiently, and productivity drops off.

Freud distinguished between fear and neurotic anxiety. Fear, he said, results from an actual danger from an external object, person, or situation. The source of neurotic anxiety, on the other hand, is an unknown danger, something vague, something not precisely definable and lacking a definite object.

A further distinction needs to be made between neurotic anxiety and the mild apprehensiveness that everyone feels from time to time. At its most basic, neurotic anxiety consists of a powerful emotion evoked when impulses from within the person are misperceived as emanating from the environment.

Anxiety, like depression, exists in a continuum. Some people rarely feel anxious, while others are regularly crippled by overwhelming and persistent fears, a disorder psychiatrists call free-floating anxiety. How are we to draw the line between "normal" anxiety and anxiety that is pathological?

In the 1960s psychologists took the first steps toward answering that question by using questionnaires, interviews, and extensive psychological, physiological, and behavioral tests. From these techniques a cluster of symptoms of the anxious person emerged. He or she lacked confidence, regularly experienced guilt or worthlessness, feared new ventures, fatigued easily, and often expressed irritability, discouragement, and uncertainty, along with varying degrees of suspicion and distrust of the motives of other people. Accompanying these symptoms was an underlying tension that in some anxious people burst forth into full-fledged anxiety attacks, in which some or all of these symptoms were greatly intensified.

These observations, mainly the work of psychologist Raymond Cattell, resulted in an important point: Popular opinion to the contrary, anxiety is not a stimulus for enhanced performance or a motivator for greater achievement. Successful and creative people are in general *less* anxious than others. Presumably their knowledge and skills counteract the tendency to become anxious. When fans asked Joe DiMaggio how he always remained so relaxed on the playing field, he answered, "Because I knew what I was doing."

Anxiety is our most pervasive emotional disorder; the response that it most commonly generates is phobia, as a means of avoiding an anxiety-creating situation. It is estimated that 19.9 million Americans suffer from these two disorders. Among women in all age groups, these disorders rank as the number-one psychiatric condition.

The most common chemical approach to treating anxiety is the self-administration of alcohol. In moderate doses alcohol may temporarily lessen anxious feelings, but it does so at the cost of additional, often heightened anxiety when the alcohol wears off. In addition, alcohol is a depressant, one of the reasons the suicide rate among alcoholics is so high. Barbiturates can bring about the same temporary lessening of anxiety, but they too are depressants and, like alcohol, over the long haul are addictive.

In the late 1960s a major effort was begun to develop a drug that could lessen anxiety without inducing sedation or risking addiction. The basis of this effort was electrophysiological research that showed that anxiety is associated with increased activity of particular brain areas. Electrical stimulation of certain nuclei in the thalamus or the limbic system regularly induces anxiety; repeated stimulation induces the same symptoms each time; and the intensity of the anxiety is directly related to the strength of the electrical current. Both human volunteers and animals were studied. Because the animals—usually rats or monkeys—react only to real danger or threats of real danger, researchers elicited anxiety equivalents by creating lesions in the ventromedial hypothalamus, an integrating area for the autonomic reactions that accompany emotions. The animals responded with hypersensitivity—the animal equivalent of anxiety—to small electric shocks, crowding, or even normal handling.

In human beings, too, anxiety can be produced or relieved by discrete surgical lesions created during needed brain operations. If the surgery destroys areas of the frontal lobes and their intimate connections, anxiety results. Destruction of specific parts of the thalamus produces relief from anxiety, probably as a result of the interruption of nerve impulses to and from the frontal lobes.

The most significant advances in understanding anxiety, however, developed not from experiments or operations but from an unlikely source and in a roundabout way. In 1945, F. M. Berger, a pharmacologist at the British Drug Houses Ltd. in London, and a collaborator, William Bradley, were attempting to synthesize new antibiotics. Penicillin had just been discovered, and although it worked marvelously in certain bacterial infections caused by gram-positive bacteria, it had no effect on infections caused by gram-negative bacteria.

Berger and Bradley started by chemically modifying a popular disinfectant that the manufacturer claimed could kill gram-negative rods that were unaffected by penicillin. Bradley suggested that slightly lengthening a carbon side chain might enhance the antibacterial effects of the substance. So it did, to a slight degree. Berger tested the new compounds in mice. "The compounds," he reported, "much to my surprise, produced reversible flaccid paralysis of the voluntary skeletal muscles unlike that I had ever seen before."

By adjusting the dosage and method of administration—by mouth or by injection—Berger discovered that he could induce muscular relaxation and seeming paralysis of all voluntary muscles in rats, mice, and other small laboratory animals. The animals went limp but remained awake and alert. Their eyes stayed open, they followed events around them, and most important, they reacted appropriately if poked, prodded, or otherwise disturbed. Their breathing, heart rate, and other autonomic activities appeared unchanged.

Animals treated with mephenesin, the most thoroughly investigated of the new compounds Berger and Bradley developed, looked and acted in ways very different from animals given anesthetics or hypnotics. At small doses—well below those needed to produce paralysis or weakness—"[t]hese compounds had a quieting effect on the demeanor of the animals. The effect was described as tranquilization," wrote Berger of his experiences in an essay, "Anxiety and the Discovery of the Tranquilizers." He continued: "When mephenesin was introduced as an agent to produce muscle relaxation during light anesthesia, it was discovered that it could allay anxiety without clouding consciousness and it could induce in tense and anxious patients the state of relaxation characteristic of normal healthy human beings."

But mephenesin was far from a perfect drug. Large doses were required for it to be effective, and it was active for only a short period of time. Berger subsequently directed his research toward developing mephenesinlike compounds that in reasonable doses were capable of producing longer-lasting paralysis of voluntary muscles. Accordingly, he considered a compound promising as an antianxiety agent if it met these two criteria: if it led to loss of the test animal's ability to right itself after rolling over, and more important, if it caused no significant excitement prior to the onset of the paralysis.

Berger worked with a new collaborator, Dr. B. J. Ludwig, and after several years their research culminated in the discovery of the drug meprobamate, subsequently given the trade name Miltown. Like Laborit, Cade, and others, Berger and his collaborators had no grand plan or knowledge about drug-receptor interactions. They simply synthesized chemicals in the laboratory and administered them to test animals and, eventually, human volunteers. The effects were noted and compared with those of other chemicals. Suitable additional modifications were then introduced.

Under the influence of meprobamate, monkeys that had formerly been vicious and difficult to handle without gloves turned tame. Electrophysiological measurements showed that the drug primarily affected the thalamus and limbic system, what Berger termed "the biologic substrate of anxiety."

When meprobamate was approved for use in humans, the patient population that responded to it was of equal interest. When a relaxed, well-adjusted, generally nonanxious person took a test dose, he or she couldn't distinguish the drug from a placebo. This was in sharp contrast to chemicals such as barbiturates and alcohol. Although these produce different effects in emotionally stable and emotionally unstable people, they bring about subjective effects in both groups. In short, emotional *in*stability—specifically, anxiety—was the psychological experience primarily affected by meprobamate. "The drug improves performance, persistence, and behavior of emotionally unstable individuals so that they, under the influence of meprobamate, react like emotionally stable individuals," said Berger.

Thus the unique and most important feature of meprobamate was its specificity: It worked only with anxious people and only affected those brain areas involved in anxiety. At the time, it was the magic bullet that Berger was seeking. Thanks to meprobamate, people who were taking strong antianxiety agents such as chlorpromazine to fight off fatigue, fear, or habituation would no longer have to cope with significant mental side effects such as difficulty in concentrating and remembering. Nor would they appear "drugged" or otherwise distinguishable from people not taking the medication. It was an exciting and momentous discovery, and it had profound implications that extended far beyond psychiatry and pharmacology.

For instance, what was the proper role for meprobamate in an era that had already been given the designation the Age of Anxiety? How much anxiety is normal, and when should an agent like meprobamate be sought by a patient and prescribed by a doctor? Such questions regarding the proper use of tranquilizing agents remain as relevant today as they were in April 1955, when meprobamate first came on the market as a prescription drug.

Typically, the criticism regarding the use of such agents goes something like this: "I want to be myself. I'm not pleased that I'm an anxious person, but at least that's me. If I take a chemical, then I'm not really dealing with the problems that are making me anxious. The drug is only covering up something that I need to discover about myself. Besides, I'm bothered with the ethics and morality of changing my attitudes not by my own effort but by means of a chemical."

Concerns like these are neither trivial nor easily countered. It is true that anxiety often serves as a signal—a word employed by psychiatrists since Freud—that something is going on in the psychic sphere that needs to be addressed. But it is also true that personal problems can be addressed only when anxiety remains below a certain level of intensity. Above this level, anxiety becomes increasingly disruptive and disorganizing. If anxiety can be chemically eased—not eliminated, but reduced to tolerable limits—then events in the psychic realm can be dealt with properly and effectively. Berger was sensitive to these issues, and at an international symposium held in April 1970, fifteen years after his discovery of meprobamate, he spoke about them eloquently, movingly, and convincingly:

> It would be wrong and naive to expect drugs to endow the mind with new insights, philosophic wisdom, or creative power. These things cannot be provided by pills or injections. Drugs can, however, eliminate obstructions and blockages that impede the proper use of the brain. Tranquilizers, by attenuating the disruptive influence of anxiety on the mind, open the way to a better and more coordinated use of the existing gifts. By doing this they are adding to happiness, human achievement, and the dignity of man.

The development of meprobamate launched the treatment of anxiety into a new phase. No longer was the chemical modification of this pervasive form of internal suffering and distress relegated to the realm of science fiction. It now appeared likely, even inevitable, that newer, stronger anxiety-relieving drugs would be forthcoming.

Indeed, in an era of pervasive tension, the quest for additional antianxiety agents took on a special urgency in light of certain practical needs. One such need was for a drug possessing a midrange potency somewhere between meprobamate and the much

stronger antipsychotic phenothiazines like chlorpromazine. Since
the antianxiety effect of meprobamate was comparatively weak,
patients often required many pills to obtain relief. This increased
the likelihood of an overdose, side effects, or an idiosyncratic reaction. But the antipsychotics at the other end of the spectrum
were too strong—a blunderbuss in situations calling for much less
powerful weapons. They also had too many potentially serious
adverse effects to warrant their use in the anxiety syndromes characteristic of neurosis.

The research process that evolved to develop new antianxiety
agents was strikingly different from earlier efforts with psychoactive substances. Cade, Kuhn, and Kline exemplify a kind of
Heroic Age, when bold and powerfully innovative individuals pro-
ceeded on the basis of intuitions and hunches. In the era that
followed, and which continues today, the bulk of research is a
combined effort by teams of scientists and technicians.

One reason for the shift away from solitary workers was the
sheer complexity of the task. In addition, new technological developments having to do with the discovery of receptors in the
1970s and 1980s ushered in an era of research in which no single
person could expect to possess the requisite knowledge or tech-
nical skills.

Competition between enterprising pharmaceutical manufacturers had already intensified by the late 1950s and early 1960s. The
potential payoffs for the development of superior psychotropic
drugs seemed limitless. Not only was there a large pool of patients
suffering from various emotional disturbances, but the development of meprobamate enlarged the target population to include
hundreds of thousands, perhaps millions, who would welcome a
safe chemical means of avoiding even perfectly normal levels of
anxiety. These people would not only be enthusiastic consumers
of anxiety-quieting agents, it was assumed; equally important, they
would be able to afford the new drugs.

Still, research teams, no matter how large or heterogeneous, are
composed of individuals, and it is individuals who select one av-

enue of research rather than another. Credit for the next development in psychoactive research belongs to a theoretical chemist who as early as the 1930s had pondered the mysteries of chemical structure and the mind.

At that time, Dr. Leo H. Sternbach, a young chemist in Cracow, Poland, was working as a research assistant on modifying and synthesizing compounds to learn their chemical nature. Theoretical chemistry is carried out on the basis of an implicit act of faith. The practitioners of this intellectually demanding discipline assume that if chemical compounds are studied with the objective of learning their properties and extending knowledge about them, some rewarding practical application will eventually emerge. But practical applications are not their principal goal. The theoretical chemist is inspired by the simple question: "I wonder what would happen if . . . "

Sternbach's research in the 1930s involved a group of chemicals known as the benzophenones, but when his best efforts didn't lead to anything particularly promising, he stopped this particular line of research and turned his attention to other matters.

In early 1954, now at the Hoffmann–La Roche laboratories in Nutley, New Jersey, Sternbach decided to resume the investigation he had abandoned in Poland and test for psychoactivity some of the compounds he had synthesized earlier. He searched the literature and found that in the intervening twenty years no additional research had been done on the compounds he had worked on in Poland. To this extent, their biological activity remained unknown.

By this time, the success of the phenothiazines and meprobamate was inspiring scientists like Sternbach to develop new and superior agents. In order to do this, the precursor chemical they used had to conform to certain specific requirements: It had to be readily available, be easily synthesized in sufficient quantities, and most important, have potential for undergoing modification into many analogues.

Sternbach added a constituent to one of his compounds, since such constituents often impart activity to substances that were inactive when tested on living organisms. In the process he discovered he had made an error in his original estimation of the compound's chemical structure. Once he corrected it, the change in chemical structure made possible forty new derivatives, but none of them evinced psychoactive properties. Discouraged, he made one last try. The resulting compound, Ro-5-0690, was then ready for testing, but a more pressing matter came up and Sternbach placed it on a shelf. A year and a half later, an associate chemist who was cleaning up the lab and discarding chemicals came upon Ro-5-0690. Since its chemical analysis was already completed, he suggested that the substance be sent on for analysis as a potential psychoactive agent.

It was by this circuitous path that "a white crystalline powder" was finally analyzed. Two months later, Dr. Lowell O. Randall, director of pharmacologic research at Hoffmann–La Roche, reported: "The substance has hypnotic, sedative, and antistrychnine effects in mice similar to meprobamate. In cats it is about twice as potent in causing muscle relaxation and ten times as potent in blocking the flexor response."

Although no one could have foreseen it at the time, these sentences would stimulate the most intense investigation in history of the effects of drugs, specifically the class of drugs known as the benzodiazepines, of which Ro-5-0690 was the first, on the human psyche.

Clinical testing of the new drug, called chlordiazepoxide, or Librium, revealed that it was more powerful than meprobamate and could relieve anxiety with a minimum of side effects.

The pace of benzodiazepine research now quickened, and in a short time fifty-five new benzodiazepines were synthesized for testing by Sternbach and colleagues. By simplifying and shortening the Librium molecule, Sternbach synthesized another compound that within a decade would become a household word, a drug so popular that the Rolling Stones would write a song about it, "Mother's Little Helper." That drug was Valium—diazepam—

and it was five times more potent than Librium. A third benzo-diazepine, oxazepam or Serax, soon followed.

For each of these successes, hundreds of other drugs proved worthless. Of fifteen hundred drugs tested in the laboratory between 1957 and 1970, only about twenty proved sufficiently promising as antianxiety agents to prompt testing with human subjects.

By now, it was clear that not every benzodiazepine was psy-chically active. The neuropharmacologist's task was to tease out by chemical manipulation those drugs that possessed the power to modify the mind. The challenge was formidable. No one could now doubt that benzodiazepines worked marvelously as anti-anxiety agents, but their mechanism of action remained obscure. It was known that they could modify the synthesis, release, and reuptake of a number of neurotransmitters, so to this extent the drugs were not a complete mystery; but how did they work at the cellular-molecular level?

In October 1976, Richard F. Squires and C. Braestrup, two young brain researchers working in a small pharmaceutical com-pany in Copenhagen, discovered specific receptors for Valium in nerve cell membranes prepared from rat brains. They did so by tagging Valium with a radioactive substance and, after injection, tracing the drug to its final destination. They discovered that the binding sites were largely restricted to the central nervous system and that the distribution was uneven: The highest concentrations were in the cerebral cortex, intermediate concentrations were in the limbic system and cerebellum, and the lowest concentrations were in the brain stem and spinal cord.

Radioactivity studies also provided a new way of estimating the antianxiety (anxiolytic) capabilities of a particular benzodiazepine. Neuroscientists would first inject radioactively tagged Valium, then follow it with a different radioactively tagged benzodiaze-pine. They found that the potency of the competing benzodiaze-pine correlated with its ability to displace Valium from the benzodiazepine receptor. The faster and more firmly the competing drug linked to the benzodiazepine receptor, the greater was its effectiveness as a tranquilizer.

Another key finding about benzodiazepine receptors soon followed. All of them seem to be coupled to receptors for the inhibitory neurotransmitter GABA (see Chapter 2), the most prevalent neurotransmitter in the brain. (It occurs at perhaps as many as one-third of all synapses.)

In essence, benzodiazepine (BZ) receptors and GABA receptors exist together in an interactive complex known as GABA-BZ. The exact molecular mechanisms whereby the benzodiazepam-GABA interactions lessen anxiety are still not precisely known, but it makes intuitive sense to associate an inhibition of anxiety with the action of an inhibitory neurotransmitter such as GABA. All the behavioral correlates of anxiety involve excitation—increased pulse, rapid breathing, and restlessness. If anxiety escalates, the generalized response is marked by hyperexcitability and panic. Drugs of the benzodiazepine family reverse all these signs and symptoms of anxiety, inhibiting the hyperexcitability response and thereby quieting the patient.

GABA is an excellent choice of neurotransmitter for bringing about such changes since it inhibits virtually all neurons when it is applied locally to their outer membranes. It does this by increasing membrane permeability to chloride ions, thus stabilizing the resting membrane potential and preventing a further spread of neuronal discharge. This allows additional chloride ions to enter the neuron and further hyperpolarize the cell membrane, reducing the cell's "excitability," that is, its tendency to "fire." This action of GABA decreases epileptic seizures and muscle spasms—inhibitory actions—so it makes a good deal of sense that it should be associated with another inhibitory action, anxiety control.

None of these findings answer an important question about the GABA-BZ receptor complex, namely, how did it evolve in the first place? Why would the human brain, whose evolution has occurred over millions of years, contain a receptor for a foreign chemical discovered in the mid-twentieth century? Since nature obviously did not conveniently outfit our brains with an appropriate receptor for Valium and its cousins in anticipation of their development, some have hypothesized that the brain contains somewhere within its mysterious folds a substance or substances very similar to the BZ drugs, a kind of natural Mother's Little Helper. But few neuroscientists believe this at present. For one thing, no such compound has turned up despite assiduous efforts to find it.

Instead, Richard Squires suggests, BZ receptors may consist of amino acid sequences that have existed within the brain for millions of years and that have become necessary for GABA-operated functions such as opening channels for chloride or closing channels for calcium. Squires's suggestion makes more sense than the theory of a "natural Valium" within the brain, but it isn't completely satisfying either.

So at this point, all one can do is contemplate the improbability of what actually happened: Valium and the other BZ drugs were discovered by a series of chemical manipulations that started with a chemist in Cracow playing around with some unknown chemicals that just happened to fit nicely onto a receptor that has cohabited for millions of years with the receptor for the brain's main inhibitory neurotransmitter. It is a strange and astonishing series of events.

CHAPTER 14
MR. GUINEA PIG

*O*pium, the solidified juice of the opium poppy, is the most ancient mind-altering substance that human beings have used.
Poppy residues can be found in Stone Age lake dwellings in
northern Italy and Switzerland. More than four thousand years
ago, the Sumerians of the Tigris-Euphrates basin, near present-
day Iraq, planted poppies. They called its juice "lucky" or
"happy"—probably the first description of opium's mood-altering
properties.

Since opium also induces sleep and relieves pain, early physicians took to administering it for various bodily ills. Its dual use
as a mood elevator and a pain reliever accounts for the ambivalent
attitude that most societies hold toward it. No substance has ever
been discovered or synthesized that is superior to the opiates
when it comes to relieving pain; nor has any substance over the
last two millennia wrought a comparable degree of social or personal havoc.

At the turn of the century, more than one percent of the pop-

ulation of the United States was addicted to opium-containing
products, chiefly in the form of patent medications. Unlike today,
people in the rural areas were particularly heavy users. In 1885
one out of every 150 people in the state of Iowa was addicted to
an opiate.

The most popular opium-containing remedy at that time was
laudanum, a potent mixture concocted in the seventeenth century
by one of the most influential physicians in the history of British
medicine, Sir Thomas Sydenham. Laudanum consisted of opium,
saffron, cloves, and cinnamon dissolved in a pint of canary wine.

The popularity of patent medicines containing opium is not
hard to understand. They tasted good, relieved pain, elevated
mood, quieted infants, induced sleep, tranquilized the anxious
and overwrought, suppressed coughs, and stopped diarrhea. Their
ready availability led to frequent and heavy ingestion. They were
cheap compared with physicians' fees; their addiction potential
was thought to be low; and most important of all, they not only
cured people of their ailments but made them feel good.

Also contributing to increased opiate usage was the availability
of opium's active ingredient, morphine, in pure form. It had been
isolated in 1805—the first psychoactive substance to be extracted
in such form from a plant—by a German pharmacist, Friedrich
Wilhelm Serturner. He named this new substance after Morpheus,
the Greek god of sleep and dreams. It retained many of the
sought-after properties of opium and was far more potent. Only
one barrier prevented its widespread use: Morphine is poorly absorbed from the stomach, rendering it ineffective when swallowed.
But that changed with the invention of the hypodermic needle
some fifty years later; morphine could now be injected as a pain
reliever. During the Civil War, large numbers of soldiers were
relieved of their pain by morphine, but they were thereafter com-
pulsively driven to repetitive use of the drug—living and tragic
proof of morphine's addictive potential.

Morphine does not relieve all pain; it is highly selective. It
works best with dull, intermittent, cramping types of pain and has

little impact on acute pain. Moreover, its principal effect is to disconnect the painful sensation from the emotional reactions that accompany it. The patient is still aware of his pain but feels detached from it, uninvolved. This observation, which dates back more than a hundred years, provides a hint about the drug's action. If morphine modifies emotional responses, it must be acting on brain areas that are concerned with the experience and expression of emotion. Indeed, suggestions that morphine and opiates in general operate in specific areas within the brain go back to 1700, when a book with the intriguing title *The Mysteries of Opium Revealed* by one John Jones suggested as much.

In the 1950s, Arnold Beckett, a professor of pharmacology in London, proposed the existence of opiate receptors. He suggested that the effectiveness of opiate drugs depended on how closely they fit these receptors. The chemical structure of morphine had been determined in 1926, and chemists were now searching for ways of modifying it so as to retain the drug's pain-killing properties but to eliminate its propensity for addiction.

The molecular configuration of these new compounds was crucial in determining their potency, as well as whether they acted as an agonist, mimicking morphine's actions, or an antagonist, blocking or reversing these actions. One substitution on the morphine molecule results in nalorphine, which exhibits either antagonist or agonist activity, depending on dosage. The same substitution, when carried out on a semisynthetic morphine derivative, oxymorphone, produces naloxone, a pure narcotic antagonist. Although naloxone differs from morphine by only a small structural modification, its antagonistic effect is powerful. One dose not only reverses morphine's pain-killing properties but precipitates an acute withdrawal reaction in addicts within seconds.

Another variant, etorphin, is so powerful an antagonist that a small drop on the tip of a dart will drop a charging elephant in its tracks. It also relieves pain in a dose one five-thousandth the effective dose of morphine. Obviously, this action could not come about as the result of overpowering great numbers of brain cells; the quantity of drug is too minute.

The intensity of the reactions that drugs like naloxone and etorphin caused strongly suggests that they are acting at specific sites. But while it is one thing to suspect the existence of a specific receptor, it is a daunting task to discover the receptor's location and mechanism of action. One stumbling block that impeded research for many years was the additional presence in the brain of *non*specific binding sites for opiates.

Like many chemicals within the body, opiates bind not only to their own specific receptors but to other receptors as well. Indeed, many different molecules on the surface of nerve cell membranes may serve as "homes" for opiates. This means that if a radioactively labeled opiate is added to a preparation composed of brain cell membranes, the opiates will "stick" to many sites that have nothing to do with pain relief, euphoria, or other opiate-induced actions.

To overcome this obstacle in their search for a specific opiate receptor, in 1972 two researchers at the Johns Hopkins University School of Medicine in Baltimore, Candace Pert and Solomon Snyder, employed a radioactive form of an opiate drug with an extremely high amount of radioactivity per molecule. This made it possible to measure exceedingly small concentrations of the drug in brain tissue. They also developed procedures for washing away opiates from the nonspecific binding sites, based on the presumption that opiates that bind to specific opiate receptors would be held more firmly than those attached to nonspecific sites, and therefore would be resistant to displacement. When you hose down a car using a very weak stream of water, not much dirt will be washed off the surface. As you increase the force of the stream, more and more dust and dirt particles cease to "bind" to the car's surface. After the most powerful stream is brought into play, only the most firmly adhering particles remain. In this analogy, the most firmly adherent opiate particles are those bound to specific opiate receptors.

In order to test the reliability of their research, Pert and Snyder homogenized rat, mice, or guinea pig brains into a "brain soup." Radioactively tagged opiates were then added to the soup. They found a dramatic correlation between an opiate's pain-relieving properties and the likelihood of its binding to specific opiate receptors. They also discovered, as did other researchers in New York and Sweden, that the number of opiate receptors varies in different parts of the neuron, with the greatest number at the synaptic membranes.

Next, Pert and Snyder aimed at pinpointing those areas within the brain itself that contained the greatest concentration of opiate receptors. They measured the binding density of radioactively labeled opiate drugs in different brain regions and found dramatic differences. The greatest concentrations were in the thalamus, a sensory integration area for all sensations except smell, and in the amygdala, the hypothalamus, and other components of the limbic system, an important region, as has been noted, for emotional expression.

These findings correlated with the euphoria generated by opium. Closer examination revealed even more precise localization. Within the thalamus, the receptors are preferentially located within the medial portion, which is the area involved in deep burning chronic pain—exactly the type of pain that opiates relieve most successfully.

The theory that the brain contains opiate receptors was now clearly confirmed. That again raised the question of why there should be binding sites in the brain for substances that are not only foreign to it but are in no way essential to life. The proportion of the world's population who ever encounter, much less use, an opiate drug in their lifetime is minuscule. Why should the brain have receptors for distillations of the opium poppy?

One intriguing possibility was that the body itself produced opiatelike substances—endorphins—with which these receptors reacted. This hypothesis seemed plausible enough to scientists to be worth pursuing, and in the 1970s several international teams fiercely competed to find these natural opiates.

In 1973 at the University of Aberdeen in Scotland, Hans Kosterlitz, a seventy-one-year-old neuroscientist with an owlish, gnomelike appearance, combined his talents with those of researcher John Hughes to identify and chemically characterize the body's opiatelike substances. They took advantage of the fact that opiates inhibit electrically induced contractions of certain smooth muscles taken from the large intestine of the guinea pig. As odd as it may sound, over the previous twenty years Kosterlitz had forged for himself an international reputation based on his observations that when dissected and strung out, the guinea pig ileum (a part of the large intestine that squeezes waste into the distal colon) responded to a mild electrical current with a series of twitches. He discovered, moreover, that the number of twitches decreased when opiates were applied. The greater the pain-killing capacity of the opiate, the greater its ability to inhibit this twitching. (Kosterlitz was known to the Aberdeen medical students as Mr. Guinea Pig, and as a result of his work, the guinea pig ileum became widely used in the neuroscience community as a quick means for testing the potency of new narcotic drugs.)

For opiates to have this inhibitory effect, Kosterlitz theorized that there had to be receptors in the ileum muscle similar to opiate receptors in the brain. So the next step was to experiment with brain tissue. Hughes obtained large quantities of pig brains from an Aberdeen slaughterhouse and carried out the arduous task of reducing them to crude extracts. Bioassays indicated that they contained opiatelike compounds. On an August afternoon in 1973, Hughes and Kosterlitz applied a few drops of the pig brain extract to another smooth muscle preparation and stimulated it electrically. Not only did the preparation cease twitching, they were even able to reverse this effect and restore normal twitching by adding the opiate antagonist naloxone to the mix.

After much additional chemical analysis—and prodded by the

specter of two American teams breathing down their necks in competitive pursuit—Kosterlitz and Hughes published in 1975 the chemical structure of the morphinelike material they had isolated from the pig brains. Dubbed enkephalin (from the Greek for "in the head"), it was a peptide consisting of two chemicals, both of them chains of five amino acids. The two molecules were identical except for the terminal amino acid on the chain; one contained leucine, the other methionine. What the two scientists had identified was a natural opiatelike substance in the brain.

This discovery (the enkephalins were subsequently renamed endorphins, a shorthand term for the brain's endogenous opiate), coupled with the earlier demonstration by Pert and Snyder of opiate receptors, initiated an intense effort by pharmaceutical companies to synthesize opiates similar to the brain's own. Success in creating the first nonaddicting opiate would be a gold mine, given human beings' eagerness for eliminating pain and experiencing pleasure. Almost two decades later, however, the results remain disappointing. A nonaddicting pain-killer as powerful as morphine still appears to be several decades away.

✛ ✛ ✛

As we have seen, radioactive tagging and receptor labeling had become indispensable means for mapping receptors in the living brain. (At this point PET scans, the most elegant visualizing technique of all, still lay in the future.) Receptor-binding techniques represented a quantum leap in our understanding of the brain. They made possible dramatically selective localizations and visualizations of receptors under the microscope. Receptor autoradiography answered questions that only a few years before had seemed too fanciful even to ask.

For instance, do opiate receptors occur only within the brain or at the spinal level as well? Receptor autoradiography revealed a dense cluster of receptors concentrated in narrow bands in the dorsal part of the spinal cord. This area, called the substantia gelatinosa, is the first way station for pain perception when injuries occur in the fingers, toes, and other peripheral structures. Opiates work at the spinal cord level to block the transmission of such pain upward to the brain. Other areas along the path from spinal cord to brain are also richly endowed with opiate receptors.

Another question is whether individuals differ in the quantity and functional integrity of their opiate receptors. Lars Terenius, of the department of pharmacology of the University of Uppsala in Sweden, an early researcher on receptor binding and an expert on autoradiography, thinks they do. "Ultimately, the endorphin systems are components of what we call personality," he says. This suggests that pain perception and sensitivity are related less to a person's "stoicism" than to the organization of opiate receptors within his or her brain and nervous system.

Even more important than their usefulness in demonstrating opiate receptors within the brain, radioactive binding techniques provided a window through which neuroscientists could observe changes in chemical activity within the brain. It also provided pharmacologists with a powerful tool for developing more potent drugs. By comparing chemical structure with chemical activity, new drugs could be developed and their site of action localized. Binding sites not only for opiates but for dopamine, serotonin, and benzodiazepines could be seen as if illuminated by a powerful search beam.

Opiate addiction involves mechanisms and brain pathways similar to those of cocaine. As we discussed in Chapter 12, the pleasure associated with these drugs appears to be mediated by the neurotransmitter dopamine, acting well below the cerebral cortex within two principal brain areas, the nucleus accumbens and the ventral tegmentum. Both of these are part of the limbic system and are known to be associated with the pleasure response. The drive to sustain this pleasure response is so powerful that the opiate, like cocaine, will be used more and more if it is available, and more and more will be needed to get the same effect. The user will rapidly become profoundly addicted and will suffer severe withdrawal symptoms upon forced or self-imposed cessation of use. It was once thought that the mechanism of opiate addiction was based on the need to stave off the frightening physical symptoms of withdrawal, but it now seems clear that it depends on the reinforcement/pleasure response.

An experiment with amphetamine showed that rats who have been newly introduced to morphine will push buttons to have the drug delivered directly to the ventral tegmental area. When their supply is cut off, signs of withdrawal fail to appear. But if the drug is delivered to another nearby brain area, the periaqueductal gray, withdrawal occurs promptly when the drug supply is stopped. Such experiments, almost surgical in their precision, distinguish brain areas involved with pleasure from those that mediate withdrawal.

One way of investigating what is happening at the chemical level is to give an animal the opiate antagonist naloxone. Within seconds after the injection, opiate effects are turned off and a severe withdrawal reaction is induced.

In the laboratory naloxone can be used as a chemical probe: It will displace opiates from those brain areas important to the addiction process. This leads to a withdrawal reaction—greater and more desperate efforts on the part of the animal to obtain additional drug. This frantic drug-seeking behavior corresponds at the chemical level to the reoccupation of the receptor sites now occupied by naloxone. But that can't occur because naloxone exerts a tremendous grip on the opiate receptor, with all the tenacity of a holder of a winning ticket waiting to cash in at the pari-mutuel window.

Nor are naloxone's effects limited to opiates. It alters the threshold for activity in the brain's pleasure centers for amphetamine and cocaine as well. This effectiveness of the same antagonist for widely different chemical substances (opiates, cocaine, amphetamine) is another indication of a common mechanism of action within the brain, a utilization of the same pathways. Other clues pointing to unity of action among these mind-altering drugs include the facts that amphetamine will increase the pain-relieving effects of morphine and that the same areas of the brain light up on PET scans when either morphine or cocaine is administered.

As Conan Kornetsky of Boston University School of Medicine notes, "This suggests that there are areas of the brain responsible for reward, and they are similar or identical for the commonly used drugs of abuse." At an international conference in New York in December 1990, Kornetsky went on to state:

> Most substances of abuse, but especially the opiates and the psychomotor stimulants like cocaine and amphetamine, share a common underlying neuronal action that is at least in part responsible for the pleasure-giving effect which in turn is responsible for their reinforcing action. At the level of the central nervous system, the action is manifested as an activation of forebrain dopamine areas of the mesocortico-limbic system. For the user this is translated into some pleasurable feelings that have been described as the high.

The nucleus accumbens is turning out to be the most important brain area involved in addiction. Nor should this be terribly surprising, since the nucleus accumbens is uniquely situated to communicate with the limbic system (which is responsible for emotional experience) and the motor cortex (which directs our actions toward seeking out pleasurable experiences). Some measure of the intensity of the feelings created by drugs can be gathered from the vocabulary employed by drug users. Rushes, hits, scores—these terms also describe intense sexual and orgiastic experiences.

So far, neuroscientists have not made up their minds as to whether opiates, cocaine, and other commonly employed mind-altering drugs exert *permanent* physical effects in the brain. There doesn't seem to be any doubt, however, that long-lasting, perhaps permanent psychological changes are induced. Indeed, if craving for a drug is a memory of only the pleasant aspects of the drug experience, not the suffering and misery, remaining off drugs is a lifetime enterprise.

Sadly, it may be a totally absorbing task as well. A former cocaine addict told me recently of the struggle she has to engage in every day in order not to return to the drug. "If I don't change myself inside—don't change basically who I am—I know I won't be able to stop myself from going back to coke." She experiences chronic alterations in her moods, her thoughts, and her ambitions for herself, she says. Like most people who have a cocaine habit, she is particularly susceptible to cravings when she is around people and in situations that remind her of her drug-using days. She finds that staying away from them requires constant effort. She attends one or two meetings a day of Narcotics Anonymous, and she speaks and reads incessantly about ways to keep off drugs. It is as if the drugs have somehow permanently altered her body chemistry, particularly the ways her brain processes pleasure. She finds that she must work continuously to keep herself stimulated, interested, and hopeful.

What could be the neurobiological basis for such profound changes in one's very self, the very core of one's personality? Behavior, brain anatomy, and brain circuitry provide indications of patterns, differing only in scale, that are encoded within the brain. Again, we are talking about levels, as in the Russian doll analogy in Chapter 8. The largest doll—the one there for all to see, like observing the obvious actions of a person—represents behavior: craving, drug-seeking, drug-taking, the "crash," the repeat of the cycle.

At the next level, the smaller doll contained within the larger one, we encounter brain sites and circuits: the ventral tegmentum, the nucleus accumbens, and other brain areas associated with pleasure and emotion.

At the third level, we encounter the molecules, the neurotransmitters and receptors that convey the chemical signals that correspond to the craving and other psychological states occurring two orders of magnitude above.

At the smallest level of all is the genetic code, the DNA that determines and influences the internal cellular processes, the interweaving of ions and molecules onto receptors and across membranes into cells where they work a kind of alchemy.

The various brain areas seem to communicate with each other in ways suggestive of musicians engaged in a jazz improvisation. While each musician plays a clearly defined role at any given moment within the context of the musical performance, their exact interplay is never precisely predictable and repeatable from one session to another.

PART III

CHAPTER 15
TRACKING THE MARIJUANA RECEPTOR

What can be expected in the future from research on the brain's receptors? In order to answer that question, I propose we look at what we have learned recently about marijuana.

Unlike most of the psychoactive chemicals discussed so far, marijuana—cannabis—is an extremely complex chemical substance and presents a challenge to even the most skilled of investigators. It is not an alkaloid—that is, it contains no nitrogen, the hallmark of other natural psychoactive compounds. Early investigators, including the famous cough-drop-producing Smith Brothers, were unable to identify its active agents.

Cannabis, derived from the hemp plant, contains more than four hundred chemicals, including simple acids, alcohols, and some sixty-one other compounds that are unique to the plant, called cannabinoids. It was not until 1964 that Δ-9-tetrahydrocannabinol (Δ-9-THC) was identified as the most active agent. Δ-9-THC is concentrated in the resin found chiefly in the flowering tops of the plant, with less in the leaves and least in the

fibrous stalks. Thus, the psychoactive potency of a cannabis preparation varies enormously depending on the part of the plant used
to make it. The purest and most powerful—prepared from the
pure resin removed from the leaves and stems—is known as
charas or, more commonly, hashish. Its concentration of Δ-9-THC
ranges between eight and fourteen percent. Next in the order of
potency—the variety most common in the United States—is
ganja, made up of dried plant material taken only from the tops
of unpollinated female plants. (The male plants are removed from
the fields prior to pollination in order to prevent the diversion of
energy and accompanying loss of potency that ensues when the
female plant invests its energy in seed production.) Known as
sinsemilla (from the Spanish *sin semilla,* "without seeds"), this
version of marijuana has a THC content of from four to eight
percent. Bhang, the weakest form, is prepared by using the re-
mainder of the plant after the top leaves have been picked, drying
it, and grinding it into a powder. This low-grade form of marijuana contains less than one percent THC.

Consistent with marijuana's chemical diversity are its many and
varied effects. In varying doses and in different individuals it se-
dates and relieves pain; in large doses it induces alterations in time
and space perception. It defies precise classification in Louis Lewin's scheme, since it can be an inebriant, euphoriant, phantasti-
cant, excitant, or hypnotic. The smorgasbord of its potential
actions and uses justifies marijuana's identity as a unique mind-
and brain-altering substance. Moreover, in contrast to many other
substances discussed in these pages, marijuana has a legitimate
usage in the treatment of medical problems such as asthma, glaucoma, and the nausea and vomiting often brought on by cancer
chemotherapy. It has also been investigated as a treatment for
epilepsy, insomnia, anorexia, loss of appetite, and high blood pressure. In response to marijuana's proven effectiveness against the
nausea of chemotherapy, the FDA in 1985 licensed Unimed, a
small drug company in Somerville, New Jersey, to produce a cap-
sule for sale to cancer patients, but so far the legal availability of

marijuana has not been extended to patients suffering from other illnesses that have been reported to benefit from its use.

Among the reasons for the FDA's concern is the still-unsettled question of whether marijuana is dependence-producing. Laboratory experiments directed at answering this question show positive results. If human volunteers are given large doses of marijuana every four hours over a ten-to-twenty-day period, within hours of the last dose they show irritability, restlessness, nausea, and vomiting. These withdrawal symptoms peak at eight hours after the last dose, then decline over the next three days.

But such symptoms are extremely rare under conditions besides laboratory experiments. This is because the doses in these experiments are higher than those used socially. Marijuana's effects also depend upon social factors such as the setting, the presence or absence of others (marijuana smoking is typically a communal rather than a solitary activity), and the smoker's previous experience with the plant. In contrast to other mind- and brain-altering substances, experienced marijuana smokers can get "high" more quickly and on less drug than do novice users. This phenomenon, known as reverse tolerance, may indicate a sensitization to the drug, but it is more probably the effect of having learned how to use it.

The first step involves pacing—drawing the smoke into the lungs and holding it there for twenty to forty seconds. Next, users have to learn to identify and control the psychological effects this produces, which they can do only after they have also learned to label the effects as pleasant. Not everyone successfully accomplishes this, which is one reason some people have panic reactions. Inexperienced users who lose their familiar perceptions of space, time, and personal identity may fear a loss of control or worry that they will remain in this strange, unfamiliar state forever. Such reactions are often treated by "talking the patient down" rather than by the use of medications—another indication of the importance of social rather than strictly neurological factors in producing the drug's effects.

It is clear that marijuana affects behavior in important ways. "There is little doubt that marijuana at common dosage levels interferes with many aspects of psychomotor and cognitive function in ways that adversely affect learning and skilled performance while high," according to Dr. Robert C. Peterson, former assistant director of research at the National Institute on Drug Abuse. "Dozens of experimental studies confirm that such cognitive processes as arithmetical reasoning, verbal and non-verbal problem solving, and short-term memory are adversely affected."

In order to discover the reasons for such alterations of mental functioning, scientists have concentrated on the drug's effects within the brain, including alterations of the nerve cell membrane, changes in the turnover of serotonin and dopamine, and variations in the synthesis of prostaglandins, a vast family of hormonelike chemicals derived from essential fatty acids. Marijuana's effects are so varied that its different actions—as a sedative, a hypnotic, a pain-killer, and a mind-altering agent—could conceivably take place via different mechanisms. For which of these were scientists to seek an explanation?

A solution to this conundrum was provided in 1988 when researchers at the National Institute of Mental Health, led by neurobiologist Miles Herkenham, developed a technique for the characterization and localization of the marijuana receptor. Before the action of any drug that binds to a receptor can be understood, the receptor's structure must be known. The story of how the marijuana receptor's structure came to be known is an enlightening example of how future research on mind- and brain-altering chemicals is likely to be carried out.

In the late 1970s two researchers, M. Ross Johnson and Lawrence Melvin of Pfizer Inc. in New York, produced a group of synthetic cannabinoids of differing potencies. One of their creations, levonantradol, was intended for use as a pain-killer. But not only did it relieve pain, it produced many marijuanalike side effects, such as dryness of the mouth, reddening of the eyes, dizziness, and mood alterations, that prevented its use as a medication. A failure in therapeutics, however, can sometimes be turned into a winner in basic laboratory research. This is what happened with levonantradol.

Kenner Rice, a chemist at the National Institute of Diabetes and Digestive and Kidney Disorders (NIDDK), created a radio-active compound from levonantradol that made it possible to track the drug and follow its path within the brain. Then Herkenham and his colleagues took this high-potency cannabinoid and used it as a probe to map the distribution of marijuana receptors within the brain. The distribution corresponded strikingly with the actions of the drug. Those areas of the brain concerned with mood, memory, and movement—the cerebellum, the basal ganglia, and the hippocampus—showed a high density of marijuana binding. In short, marijuana receptors are located in those parts of the brain that mediate the drug's behavioral, emotional, and cognitive actions.

While confirming the existence of a marijuana receptor was a major step in our understanding of the drug's actions within the brain, it remained to account for how marijuana works and what is happening at the molecular level.

At about the same time that Herkenham's work was going on, Allyn Howlett, a researcher at St. Louis University Medical School, was turning her attention to marijuana's chemical-molecular mode of action. She found that cannabinoids inhibit the action of adenylate cyclase, an enzyme that stimulates a wave of interrelated chemical influences in what neuroscientists refer to as "cascade" reactions. In most instances the end result is an amplification of the original signal as more and more chemicals within the neuron are called into play. We will encounter reactions like these on other occasions in our survey of the brain's chemical terrain since they play a prominent role in brain activity.

Briefly, this is how a cascade reaction occurs: A neurotransmitter occupies its receptor on the postsynaptic cell; it then stimulates a series of chemical reactions within the cell by activating a network of interacting intracellular messenger systems. Its short-term actions include altering the sensitivities of the receptors and ion channels to various synaptic inputs from other neurons. Other actions alter the genes and thereby effect longer-term alterations in the neuron. But however short or long the time frame, the basic mechanism is the same: The initial binding of the neurotransmitter to its receptor is amplified to include a cascade of chemical interactions.

Howlett found that adenylate cyclase stimulates the synthesis of cyclic AMP (adenosine monophosphate), a compound that sets off the cascade of reactions within neurons. She also found that the degree to which a synthetic cannabinoid inhibits this action of adenylate cyclase correlates with the cannabinoid's potency. For instance, when the powerful cannabinoid levonantradol was used, the degree of adenylate cyclase inhibition turned out to be extremely high.

In summary, this is where things stood in the late 1980s: A marijuana receptor had been identified, its locations specified, and its mechanism of action explained on the chemical-molecular level. This was an impressive series of advances considering that as recently as a decade ago the drug was known only as a substance capable of creating a high. But in August 1990 researchers at NIMH announced an even more exciting advance: the cloning of the marijuana receptor.

Paradoxically, this achievement did not follow directly from the work of either Herkenham or Howlett. It resulted from yet a third investigative effort.

In 1987, Lisa Matsuda, a molecular biologist working in the laboratory of neurobiologist Tom Bonner at NIMH, was trying to clone the genes for two substances concerned with the modulation of pain impulses within the brain: neuromedin and Substance P. She and Bonner elected to work with a similar substance, Substance K, the gene for which had recently been successfully cloned by researchers in Japan. Since Substance K is structurally and functionally similar to Substance P, the two researchers reasoned that the receptors and the genes for their formation should also be similar.

As a first step, they made a gene probe, a synthetic DNA sequence capable of recognizing and binding to a segment of the Substance K gene. The probe worked as planned, pulling out a gene that coded for a protein that had the general features of a receptor. But a receptor for what? Tests showed that it wasn't a receptor for the substances that modulated pain. Moreover, the receptor failed to provide any clues about what chemical or chemicals it would actually bind to.

Under these circumstances the NIMH group had little choice but to embark on a fishing expedition: It had to screen a large number of unrelated neurotransmitters and hormones in search of the one that would interact with the protein product of their so-far-unidentified gene. They pursued this fruitlessly for a year and a half, then tried a different approach: They mapped the areas of rat brain where the mystery gene appeared to be most active. Unknown to Matsuda and Bonner, these brain areas corresponded to the brain areas that Miles Herkenham had mapped for the marijuana receptor. This is particularly ironic, besides illustrating the specialization and fragmentation that sometimes exist in neurobiological research, because Miles Herkenham's laboratory was just down the hall from where much of Matsuda's research was carried out. When she learned of Herkenham's map of the marijuana receptors, Matsuda walked down to his office, and the two of them compared maps. In an instant Matsuda had the identity of the receptor gene that she and her colleagues had cloned. The marijuana receptor concentrations on Herkenham's map were highest in the same brain sections where the mystery gene was most active.

The next step was obvious: to test whether their gene encoded for the marijuana receptor. First, the Matsuda team placed cannabinoid drugs of varying potency in a culture of cells containing marijuana receptors and the gene. The drugs reacted with the receptors. Even more interesting, when the cloned gene was introduced into a culture of cells from the ovary of a test animal, a Chinese hamster, and cannabinoids were then put into the culture, adenylate cyclase was inhibited only in the cells carrying the transferred receptor gene. Further, Matsuda and her group confirmed Howlett's finding that the stronger the cannabinoid administered—the greater its capacity to produce highs—the greater the inhibition of adenylate cyclase.

Matsuda could now confidently announce that she and her colleagues had discovered the marijuana gene, but there was still no knowledge of the identity of the natural substance or substances that the gene coded and the body produced that bound to the marijuana receptor. Marijuana receptors in the brain are one hundred times more common than receptors for morphine and other opiates. Could it be that the euphoriant and medicinal qualities of marijuana—its ability to influence asthma and insomnia and anorexia—are based on a chemical similarity between the drug and such natural substances?

In December 1992, William Devane and Raphael Mechoulam of the Hebrew University and their colleagues identified a natural brain molecule that binds to the marijuana receptor. Dubbed anandamide, from the Sanskrit word for "internal bliss," the substance mimics the action of THC. Interestingly, it doesn't look anything like THC, but this is not of great concern to neuroscientists since a similar situation exists with heroin and cocaine, two opiate drugs that possess chemical structures far different from the body's natural opiates. The important point is not that the chemicals be structurally identical but that they bind to a common receptor.

With the discovery of the internal equivalent of THC, neuroscientists are now in a position to synthetically produce increased quantities of it through cloning. The substance can then be tested in various dosages as a medicine for the illnesses and conditions that are sometimes helped by marijuana. Indeed, the multiplicity of bodily effects produced by marijuana argue for the existence of more than one marijuana receptor.

Anandamide is the only natural marijuana receptor–stimulating chemical discovered thus far, but scientists strongly suspect the existence of others. Future research is likely to involve the use of computers to design new and more specific synthetic analogues of marijuana. There are several reasons why this strategy is likely to be effective. For one, the natural marijuanalike substance (or substances) in the body can be expected to exert such powerful effects on so many different parts of the body at once if administered therapeutically that in many cases it could not be used. A person with low blood pressure, for instance, would not be able to employ the substance for her glaucoma because further lowering of her blood pressure could induce a stroke. What is wanted are ways of mimicking what takes place in the body: turning on only those aspects of the chemical's actions that would be useful in the treatment of a specific illness.

Yet another approach involves the development of marijuana receptor blockers. Marijuana is smoked by AIDS patients taking AZT both to block nausea and to stimulate appetite. It is also currently undergoing trials in the treatment of epilepsy and Huntington's chorea. In each instance, researchers are attempting to take advantage of marijuana's positive effects while selectively blocking receptors for the harmful ones.

This goal is common to the design of most new drugs. Indeed, neuroscientists working toward these objectives frequently employ terms like *tickling* or *sculpting* or *designing* their brain-altering molecules. They mean *designing* quite literally. Like dress designers who create new fashions by taking inspiration from something that has already proved a success, rather than starting from scratch, brain researchers start with a substance that has been effective and then try to improve on it.

Natural compounds derived from plants have often been the basis for the design process. Once pharmaceutical companies are in possession of a natural compound with known psychic actions, their scientists test it at random in order to discover new uses. They study its active components in animals and later in human volunteers in order to find out which component is most active. Then, like dress designers, they modify the design in search of variants that will have different or more desired effects.

They may change the most active compound to give it additional properties that will broaden its usefulness. Or they may extend the period of the drug's action so that a patient need take less of it. They may try to eliminate undesirable side effects such as nausea so that a drug that previously had to be injected can be taken by mouth. Pharmaceutical companies can point to many successes in this somewhat hit-or-miss approach to new drug design, but it is a costly and time-consuming method. Thousands of compounds may have to be tried out until someone, or more usually some team, comes up with a winner. This laborious research and development or R&D process is one of the reasons drugs are so expensive. All of us are helping to pay for the untold numbers of false leads and initially promising compounds that didn't turn out as expected.

Within the last decade, researchers have come up with ways of improving on this approach by utilizing the latest developments in biochemistry, molecular biology, and genetics.

A dress designer, as a first step, will learn everything possible about the market she is trying to reach. She will want to learn her customers' disposable income, how much they are traditionally prepared to spend on clothes, and their interests in and positions on social and political issues. Once all these factors have been taken into account, she can begin her search for the next season's new creation.

Designers of new drugs follow a similar course. First they try to understand in minute detail the target that the drug must seek out in the body. This target can be considered at the purely descriptive level, such as relaxation or tranquilization. But as psychopharmacology has progressed, the emphasis has increasingly shifted from the purely descriptive to the biochemical and molecular.

By modifying a drug's structure, studying how it interacts with

its target, and then seeking to improve this drug-receptor interaction, psychopharmacologists employ what has come to be
known in the pharmaceutical industry as a "design cycle."

As a first and most critical step, investigators must discover in
detail the structure and chemical properties of the drug *and* its
receptor. This knowledge can then be used to suggest chemical
modifications of either. The neuropharmacologist must first isolate
the receptor that the drug targets. This is usually a large molecule,
most often a protein; in many instances that protein is an enzyme.
Enzymes, as we know, speed up the splitting apart or stitching
together of molecules involved in the consumption or production
of energy; without enzymes the body's reactions would grind to
a halt.

But not all enzymes carry out helpful or beneficial activities.
For instance, the protein renin is an enzyme secreted by the kidneys that increases blood pressure. With the aid of computers,
scientists recently came up with the three-dimensional structure
of renin. They then developed analogues or variants of the renin
receptor that were designed to resist the attack of renin and
thereby prevent the elevation of blood pressure. Unfortunately,
the first analogues of the renin receptor were impractical: Some
were too large to be absorbed from the intestine, so they were
digested and rendered useless. New variants had to be developed
that would resist chemical breakdown in the intestine and, once
absorbed into the bloodstream, remain in the circulation long
enough to do a highly specific job: act as new receptors to which
renin would bind and neutralize it—inhibit its action—before it
adhered to its usual receptors in the body and thereby raised the
blood pressure.

Although renin is not a mind- or brain-altering substance, its
story serves as a model for what psychopharmacologists must do
to develop drugs using a receptor-based design cycle. Once they
know the structure and chemical properties of both the drug and
its receptor, they can make modifications in a drug that will cause
the receptor to accept it. Discovering these structures and chem-

ical properties is a complex process, however, because the brain presents the researcher with unique difficulties. Its receptors are located within the membrane enclosing the nerve cell, so they are hard to isolate and characterize chemically. In order to get around this difficulty, scientists have developed ways of breaking up the membrane and separating its components according to such factors as weight and electrical charge. Once they have isolated the receptors from the other components of the nerve cell membrane, the laborious process of identifying their chemical structures and three-dimensional shapes begins.

But knowing the structure of a receptor within the brain is not enough. It is also necessary to know how and why the neurotransmitter recognizes and interacts with that receptor. Without explicit knowledge of this process, neurochemists cannot even begin to design a drug that might interrupt or interfere with it. New technological advances such as X-ray analysis and computer graphics provide insights into chemical structures, but the configurations of most of the complex protein molecules, the neurotransmitters, remain a challenge.

"If we can define the three-dimensional structure of a drug target and create a new model of how a drug interacts with it, we can try to improve the drug," says Tom Blundell, honorary director of the Imperial Cancer Research Fund Unit of Structural Molecular Biology in London. "For example, we might suggest changing the drug molecule so that it fits better into the target. This may involve changing its electronic charge, its oiliness, or simply its shape. We then make a new molecule that incorporates all the improvements and test it to see how well it binds to the target. Usually, these studies reveal that there are further improvements to be made, so we go round the same cycle again: We synthesize a new molecule with the changes suggested, test its binding, and study the structure of the complex again. To make a new drug we must go round the design cycle many times, hoping to make a small improvement in the molecule each time."

Considerations of safety are critical at every step of the way.

Recall the tragic results when David, the amateur chemist in Bethesda, miscalculated in preparing a chemical almost identical to meperidine and inflicted irreversible damage to his brain. The qualifier *almost* says it all. In the challenging but perilous endeavor of designing drugs to alter brain function, to miss by an inch is to miss by a mile.

As the biochemical understanding of transmitters and receptors increases, the pharmaceutical industry will be seeking drugs that modify a wide range of conditions, from mild disturbances of mood and thought to serious disorders characterized by severely impaired functioning. Much of this new pharmacology will take advantage of what has been learned from technological advances. For instance, the PET scans of some depressed patients show abnormalities within parts of the prefrontal cortex, the limbic system, and subcortical brain regions. This finding suggests the existence of a functional neuroanatomic circuit involved in depression. Mood changes in individuals who have never suffered from depression are also accompanied by differences in blood flow in parts of the frontal cortex. If normal healthy volunteers merely think sad thoughts while undergoing PET scanning, the scan will show activation of these frontal areas. Thus, neuroscientists are closing in on the brain mechanisms involved in both normal and disturbed mood states.

Memory is another area in which advances can be expected. Severe memory loss is devastating. Rob a person of her memories, and you deprive her of her identity. One of the saddest aspects of Alzheimer's disease is the victims' inability to recognize those who have been close to them, to live in a bewildering world inhabited by strangers. Might the development of new drugs restore the sense of identity in Alzheimer's patients if given early in the course of the disease? Within the next few years drugs designed to enhance memory will almost certainly be developed that promise to be useful in the management of the severe disturbances associated with illnesses such as Alzheimer's. The nicotine receptor undoubtedly plays a role in normal brain functioning, particularly in memory and other cognitive processes. As described in Chapter 10, the use of nicotine increases the number of nicotinic receptors within the brain. Perhaps new and better analogues of nicotine will become available that will stimulate these receptors and lead to an increase in their number without desensitizing them, as does nicotine itself. More complete knowledge of their operation should prove helpful in the design of new drugs to enhance the receptors' actions.

Agents will also be available soon to help with the forgetting that accompanies aging. Treatments here are certain to be more controversial, however, because some loss of memory function is part of normal aging. On standardized memory testing, fifty percent of individuals over fifty years of age fall at least one standard deviation below the mean memory score for twenty-six-year-olds. This age-associated memory impairment (AAMI), as it is called, hardly qualifies as a disease—or does it? Glaxo, a pharmaceutical company that is currently heavily committed to the development of a memory drug, regards AAMI as "a disease worth treating." A critic of this approach to normal forgetting describes memory pills for normal individuals as "a commercial concept; a designer disorder for designer drugs."

In mental disorders, altering brain chemistry can certainly offer great new opportunities to help those who, because of their illness, are sometimes not capable of helping themselves. We have seen this happen over the past three decades, and drugs now under development promise even greater help. But brain-altering drugs can also be expected to help certain subtle disorders that are not considered out of the ordinary.

Adult forms of hyperactivity with their accompanying restlessness are one example. As a lifetime sufferer from this condition, I can attest to the problems and frustration it causes whenever I am forced by circumstances to attend a concert. However compelling the performance, I find it almost impossible to remain in my seat for more than half an hour, and I must go out and pace around the lobby. A suitable drug that will cut back on this unwanted motor output is not far in the future. Doctors have already discovered that opiates will greatly improve the so-called "restless legs" syndrome, in which the patient periodically experiences the nearly overpowering need to move his legs. Obviously the risk of addiction overshadows the beneficial effect that opiates may exert on this comparatively benign condition, so the goal is to find equivalent drugs without addictive potential.

Another area in which mind- and brain-altering drugs can be expected to come into common use is the treatment of sexual disorders, such as psychologically based impotence. Urologists and psychiatrists are already employing serotonin reuptake blockers to help men obtain and maintain erections. It is thought that these drugs exert their effect at the lower spinal cord, as well as in certain brain areas not yet pinpointed but that are not involved in mood. Neuroscientists deduce this because these blockers are antidepressants. The antidepressant response takes several weeks to develop, whereas psychogenic impotence can be treated from day to day (or more frequently, from night to night), with improvement occurring within an hour or two of taking the drug.

All these drugs affect mind, brain, and body in powerful ways. Some of these ways are known; some can be guessed at; others may be lurking of which we have no inkling. Drugs are powerful substances, and even those that have been properly developed and tested and have been proved to be effective for the conditions for which they are prescribed, may have adverse effects with prolonged use. A condition called tardive dyskinesia is the most conspicuous example of unanticipated harmful effects from the use of mind- and brain-altering agents. Most commonly, the illness is confined to the tongue and lips: The patient can't seem to keep his tongue in his mouth; it darts in and out constantly in a rapid, repetitive play that cannot be imitated for more than a few seconds by anyone without the illness. There may also be bizarre body movements, such as short, quick, uncoordinated jerks of the face, trunk, arms, and legs. Sometimes the afflicted person's walking is affected; the gait takes on a dancing pixie quality. These manifestations can be so grotesque and conspicuous that had the condition been seen in earlier times, it might well have been taken as a sign of some kind of spirit infestation.

In fact, tardive dyskinesia results from prolonged use of the stronger tranquilizing agents employed in the management of serious mental illnesses—neuroleptics like thorazine that work by blocking the dopamine receptors on the postsynaptic membrane of the receiving neuron. Although the explanations for tardive dyskinesia are still not entirely satisfactory, most neuro-scientists believe it results from the long-term effects of the drugs on the postsynaptic receptor. Dopamine activity is suppressed, and after a while, for reasons not completely understood, the re-ceptors become hypersensitive to *normal* levels of the neurotrans-mitter. They tend to bind these molecules with unusual avidity. As a result, motor movements become destabilized, incessant, and bizarre. Even small amounts of naturally available or administered dopamine dramatically accentuate these movements.

At the moment there is no uniformly successful treatment for tardive dyskinesia; from ten to twenty-five percent of those af-flicted retain their symptoms no matter what medications are tried. Ironically, the patients often improve temporarily when neu-roleptics are reintroduced at higher doses. It is speculated that the improvement results from the drug's ability to "clasp the neu-ron" once again at the higher dose, thus damping down *all* motor activity, including the abnormal movements. But such drug-induced improvements are only temporary and lead to a vicious spiral of recurrent and increasingly severe symptoms with each dosage increase.

From the tardive dyskinesia experience neuroscientists have learned an unsettling lesson: Not only can brain-altering drugs sometimes cause permanent harm, they may act like a time bomb that goes off a decade or more after exposure to them. Could something similar happen with the milder tranquilizers and anti-depressant agents that are now used by millions of people? What about medications like Ritalin that are now almost routinely prescribed to control hyperactivity in children? This latter possibility is particularly troublesome since a child's brain is still undergoing development and may thus be at even greater risk. Unfortunately, neither neuroscientists nor anyone else can provide reassurance about the long-term safety of present or future mind- and brain-altering drugs.

As the potential for developing a wide range of more precisely targeted drugs grows, ethical and societal questions about what might be considered intrusions into the normal human condition proliferate as well. Most of the new drugs will be aimed not so much at "patients" as at people who are already functioning on a high level. How should we view altering the brain function of a person who wishes to enhance perfectly satisfactory performance to achieve even more? What attitude should we take toward the seventy-hour-a-week workaholic who wants a brain-altering chemical in order to work even longer and harder? In the latter case there is likely to be some legitimate controversy, since many behaviors that presently fit within the category of "normal," like workaholism, are actually indicative of a deep-seated psychopathology. Workaholics neglect spouses, children, and community and social obligations in the service of total immersion in work. Psychiatrists report seeing many more such disorders of behavior and character over the last decade than used to be the case.

Similar questions are raised by the potential overuse of mind-altering substances to extend one's energies beyond exhaustion. All of us do this on occasion, and it is perfectly normal. But what about the person who consistently drives himself to a level of physical or emotional collapse? Should drugs be made available to this person to make such arduous efforts possible?

I choose this example for two reasons. First, we are likely to have in the near future drugs capable of combating fatigue and enhancing alertness without inducing addiction or dependence. The early precursors of these drugs are already in use, and at least in some people don't seem to be doing any obvious harm. Second, there is a ready demand and market poised to use such energy-enhancing drugs from the moment they are introduced. The many of us who are ambitious and tend toward workaholism would find safe, energy-conferring drugs to be of enormous appeal. But such drugs have the potential to bring about a lessening, rather than an enhancement, of mental health. Ignoring or cutting off inner feelings of fatigue breaks up the normal feedback loop that underlies good health. Under ordinary conditions, we discover what's going on within ourselves by "listening to our bodies." When we are in need of food, we start to feel hungry. Over-exertion works in the same way; fatigue is a signal that we should slow down or cut back lest we suffer some sort of physical or mental breakdown. If we cut off these internal signals in the interest of getting more and more work accomplished, we will literally lose touch with those internal sensations that guard our general health.

Other targets of new drugs will be comparatively minor problems in living, not psychiatric illnesses, including the milder forms of panic, social withdrawal—in people who can force themselves to socialize but tend to avoid others and prefer to be alone—and other socially confining rather than disabling behavioral and emotional disturbances. While it may be extreme to claim that people will soon be able to select the personality traits and qualities that they desire, it is no exaggeration to state that mind- and brain-altering drugs will be used to equilibrate moods, enhance cognitive functions such as memory, and to feel generally good about life. Some of the newer drugs already on the market hint at what is to come. Prozac, the most successful and widely employed anti-depressant in history, is being increasingly requested by people who have no diagnosable emotional illness but who feel that "something is missing" from their lives. When asked to elaborate, such people speak of a vague sense of lack of fulfillment. Many people with this experience report feeling better after starting Prozac. But obviously there is a danger in dependence on easy fixes for easy problems—it may invite a slide over to dependence on riskier fixes for harder problems. It is too easy to forget that everyday adjustment problems are part of being alive.

In the 1940s and 1950s women took amphetamines to control their appetite and retain the slim figure that was the considered ideal at the time. Many of them became dependent on the drugs. Today the epidemic of anorexia and other eating disorders virtually guarantees a ready market for new kinds of "diet pills." The most sophisticated agents in this area are almost guaranteed to be brain-altering—perhaps targeting the brain's satiety center—which raises the question of whether an obsessive desire for thinness justifies the regular consumption of chemicals that might turn out in the long run to be harmful.

An unhealthy trade-off often exists between the neuroticism of a society and the drugs that the society favors. Today the emphasis is on thinness, quickness, efficiency, self-fulfillment, driving ambition, the early achievement of career goals, the absence of inhibitions, and a certain amount of ruthlessness. Will drugs be developed that cater to these desires? If past events are any indication, they almost certainly will. In every age chemicals are discovered or synthesized and used to help people achieve whatever version of the good life prevails at the moment. To this extent, the drugs developed within any society reflect and amplify the ideals and goals of that society.

CHAPTER 16
DESIGNER BRAIN

*I*n the near future neuroscientists will be able to construct a map of an individual's brain showing its receptors and the substances that bind to them. These maps will provide a more scientific understanding of emotional and neurological illnesses. For some mental illnesses, the maps will help determine therapy. The choice of specific medications for a depressed patient, for instance, may be determined when a neuropsychiatrist looks at his brain map, identifies a deficiency, and decides, for instance, that a depression associated with low levels of serotonin may best be treated with a serotonin reuptake blocker like Prozac.

"In the future we'll have other examples of drugs [besides Prozac] aimed at one or another of the neurotransmitters," says Johns Hopkins psychiatrist and neuroscientist Solomon Snyder. Until the advent of Prozac, he points out, most of the available tranquilizers and antidepressants were modifications of agents developed between 1955 and 1960. He believes that a new generation of more potent, safer agents will be available by the turn of the century.

Another avenue of research will involve what neuroscientists call second messengers. Few neurotransmitters actually enter their target cells to bring about the alterations in activity associated with such processes as thinking, feeling, and remembering. Instead, most of them employ intermediaries.

A neurotransmitter first binds to its receptor on the outer surface of the nerve cell membrane. The receptor, which extends from the surface to the interior of the cell, then relays the information via one or more intermediaries. The first intermediary is from a family of structurally related proteins known as G proteins, which direct the flow of information from the neurotransmitter-receptor interaction on the outer cell membrane inward to the rest of the neuron. They do this by acting on a second intermediary, also found along the inner surface of the membrane, called an effector. Typically the effector is an enzyme that converts an inactive molecule within the neuron into an active one—the second messenger. It is this activated molecule that triggers the series of chemical reactions within the nerve cells, the cascades described in Chapter 15, that form the basis for the complex information processing carried out by the brain.

Since G proteins receive and convey information from many different neurotransmitter-receptor interactions, they provide a means of integrating vast amounts of information within the brain. For example, the G protein may respond to multiple receptors and then influence a comparatively small number of effectors. In this way different neurotransmitters may bring about similar actions within a neuron. Conversely, the same neurotransmitter may bring about different effects by activating G proteins, usually through different receptors.

Imagine turning the key in the ignition of a powerful car. This key-to-lock combination initiates a series of events within the car's engine. Fuel is injected, gears can be shifted, and electrical power is distributed to control lights, radio, and air conditioning. In order to bring about these different responses, it is not necessary to insert separate keys. One key does it all, by amplification and signal transduction. This is also what happens within nerve cells, thanks to the activity of second messengers.

In the literature on the action of G proteins and second messengers, the word *cascade* keeps turning up, and it is apt. "Signal amplification cascades have the characteristic property that they can be triggered by small barely detectable changes in the actual concentrations of second messengers," says Dr. Ernst J. M. Helmreich of the department of physiological chemistry of the Würzburg Medical School in Germany.

G proteins and second messenger systems hold promise for revolutionizing our understanding of both the normal and the dysfunctional brain. Abnormalities in G protein function have already been identified in alcoholism and opiate use, and they may play a role in schizophrenia and depression. One way in which drugs aimed at depression and psychosis may exert their effects is by altering G proteins or second messenger systems. Recent evidence suggests that lithium's antidepressant and mood-stabilizing effects are related to its actions on second messenger–generating systems.

With increasing knowledge, neuroscientists will synthesize drugs capable of reacting selectively with specific types of G proteins to modify malfunctioning cells. Ultimately, they will develop a wiring diagram of the neuronal membrane. Writing in *Scientific American,* Maurine E. Linder and Alfred G. Gilman of the University of Texas Southwestern Medical Center in Dallas suggest that "neuroscientists will know how receptors, G proteins, and effectors are connected. And they will be able to predict how the cells will operate in response to any combination of signals." For scientists intent on developing drug therapies, write Linder and Gilman, such discoveries will be like giving a thief a wiring diagram to the alarm system of a bank.

Another promising avenue of research is based on what Miles Herkenham of NIMH refers to as "mismatches" between neurotransmitters and their receptors. Traditionally, the sites of release of transmitters and the receptors to which they bind have been thought to lie near each other. But recent investigation shows that this is not always the case.

Herkenham suggests that this organizational independence may reflect two modes of communication used by transmitters and receptors. According to this new model of brain functioning, synaptic connections between neurons may not be as important as chemical interactions taking place over huge (for the brain) distances. Thus a transmitter may not only act at a synapse but may influence action at great distances away by simple diffusion. Controversy exists concerning the distances that neurotransmitters can diffuse after release; in many instances they are released relatively far from neurons known to have receptors for them. This is an important area for further investigation, since small messenger molecules such as peptides are thought to play a crucial role in the more subtle forms of behavior. They do this by modifying the effects of the larger, classical neurotransmitters like acetylcholine, norepinephrine, and serotonin.

This modulatory function at the chemical level may be mirrored on the behavioral level. For instance, not all feelings of joy or sadness are equally intense. These and other subtle nuances in mood and behavior may take their origin from these "mismatches" between transmitter and receptor location.

While the concept of such a mismatch is revolutionary, it is less surprising given how little we actually know about chemical transmission along the major pathways within the brain. As Herkenham and others point out, many of the major chemical pathways within the brain remain mysterious; in some neurotransmitter-receptor interactions, we are often not certain which neurotransmitter is the operative one. In other instances, we are not sure if a chemical is functioning as a neurotransmitter or if nothing more than an inactive building block is being used by the brain cell to construct a protein.

If Herkenham is right, information flow within the brain that occurs by means other than the synapses—called parasynaptic communication—may play a greater role in brain functioning than anyone ever imagined. Consider, for instance, the impact that substances carried within the blood can exert on brain function. Angiotensin, a potent chemical responsible for constricting blood vessels, acts on angiotensin receptors on the nerve cell membrane. But before it can do this, the kidney secretion renin must form angiotensin by converting the circulating precursor angiotensinogen. In this example, the kidneys influence the action of the brain by modifying events happening at the nerve cell membrane. Action at the membrane is then coupled to the activity of second messengers, which work within the nerve cell to alter the cell's metabolism or structure.

"Action at a distance according to chemical rather than anatomical determinants—this is the wave of the future in the neurosciences. This is a new way of thinking about the brain," says Herkenham. Nor should this be surprising, since chemical influence exerted over great distances is the rule elsewhere in the body. (Insulin, thyroid, and other endocrine hormones exert effects on every cell in the body.) All of which raises an intriguing question: Are we too rigid in our concepts of what is and is not brain and, by implication, what is and is not mind?

Over the past decade many peptides and other transmitter chemicals within the brain have been discovered in various locations throughout the body. These regulatory hormones, with names like Substance P (the substance that started Lisa Matsuda on the path to discovering the marijuana gene), gastrin, and lipotropin, occur not only in the brain and spinal cord but in the gut, adrenals, and sexual organs. This chemical equivalence probably forms the basis for the influence of mental processes on brain function and vice versa. Our "gut" feelings are more than mere metaphor, since some of the same neurotransmitters and peptide modulators occur both in the brain and in the gastrointestinal tract. Indeed, the mental and the physical, the mind, the brain, and the body, are intrinsically linked by means of these chemicals.

From this observation comes a profound insight into the nature of the brain. "Central to the paradigm that the mind is modulated by hormones is the recognition that the stuff of thought is not caged in the brain but is scattered all over the body; regulatory hormones are ubiquitous," says neurosurgeon Richard Bergland, author of *The Fabric of Mind*. "There is little doubt now that the brain is a gland; it produces hormones, it has hormone receptors, it is bathed in hormones, hormones run up and down the fibers on individual nerves, and every activity that the brain is engaged in involves hormones."

If Bergland is correct—and more and more neuroscientists sub-scribe to the hormonal paradigm of brain functioning—then it is likely that drugs will be developed aimed at modifying hormones and the endocrine system. Efforts to explain and modify the brain will depend on such chemical manipulations. Many bodily func-tions like sleep and sexual activity are influenced by multiple reg-ulatory hormones; so are many mind and brain activities. Thus memory, which is a function of mind and brain, is modified by acetylcholine, adrenaline, noradrenaline, and probably serotonin. No single one of these transmitters—regulatory chemicals—is spe-cifically and solely responsible for memory. Rather, all of them, together with other regulatory chemicals that are almost certain to be revealed in the future, interact with one another like the instruments in a symphony orchestra. In short, one aspect of mood or behavior may be linked to more than one regulatory chemical, and a single brain chemical may be associated with more than one mood or brain activity.

Despite the seemingly infinite number of combinations made possible by such an arrangement, neuroscientists can take comfort from the fact that there does seem to be an upper limit on the number of regulatory chemicals within the brain.

"When neuroscientists first started isolating more and more peptides several years ago, many of us began to fear that maybe we were headed for the same thing physicists encountered in their search for elementary particles," says Herkenham. "But fortu-nately things aren't turning out that way. We probably haven't found all of the brain's regulatory chemicals. But we don't expect that the number is likely to be doubled. We're confident that we've discovered most of them."

Not all neuroscientists agree with Herkenham's last point. Nonetheless, it is likely that the number of regulatory chemicals within the brain is controlled by the same considerations that limit the number of letters in an alphabet, as with the amino acid sequences described in Chapter 8. Combinations of just twenty-six letters have made possible the expression of our loftiest philosophical speculations, including "What is the relationship of the mind to the brain?" The addition of one or more letters would lead to a decrease rather than an increase of information in that sentence. Typographical errors, such as the introduction of inappropriate letters in a word, bring about confusion, not knowledge. In a similar process, the sixty or so currently known neurotransmitters and regulatory chemicals provide more than enough components for an infinitude of possible combinations.

Space-age technologies will allow for the development of synthetic analogues of the brain's hormones and chemicals. Although such analogues will still be referred to as drugs, they will ever more closely resemble the structure and function of the brain's own chemicals. Some will modify genetic expression, turning on and off certain genetic programs within the chromosomes. For instance, applications of recombinant DNA techniques will open up new possibilities of detecting and modifying the genes responsible for illnesses like Huntington's chorea (already located on chromosome four), as well as certain forms of depression like the chromosome-associated depression recently discovered to be linked with chromosome eleven.

What's ahead—in fact, what is already occurring, as we have seen—are efforts to "sculpt" chemicals that have known psychoactive effects. In this way, as in the recent work with the marijuana receptor discussed in Chapter 15, the undesirable effects of a drug can be eliminated while its good ones are maintained.

As we have noted, designer chemists must know both the structure of a receptor and how a drug interacts with it; only then can they produce drugs that might enhance or interfere with this process. Techniques like X-ray analysis and computer graphics now enable them to accomplish in an afternoon what formerly took months to do: describe the shapes of proposed drugs and picture them in three dimensions.

A particularly fascinating area of research involves the changes in size that proteins routinely undergo at the molecular level. Like all matter at room temperature, the protein molecules that make up the brain exhibit vibrating motions due to thermal energy. When large groups of atoms in a brain protein move in unison, physicists refer to the process as the "breathing" of the protein molecule. Thus a complete description of a protein includes not only its static three-dimensional structure but a description of its fluctuations over minute time dimensions, usually in picoseconds. (One picosecond equals 0.000 000 000 001 seconds.) These "breathing motions" play a major role in the function of the proteins and affect such processes as the binding of neurotransmitters to their receptors. The discovery of these breathing motions has eliminated any doubt that the brain is a dynamic organ even at the molecular and submolecular level. What is needed now are neuroscientists knowledgeable enough to apply to the human brain new insights from the world of physics.

From its beginnings in the first three decades of this century, quantum mechanics held great promise as a means of learning the deeper mysteries of biology. "The eventual addition of biological concepts to quantum mechanics is a foregone conclusion," Niels Bohr wrote sixty years ago. Such optimism seems less brash today. Studies of neurotransmitters and receptors reveal that events within the brain involve a chemical dialogue, a conversation in which chemicals talk to each other. This molecular cross-talk occurs at receptors, but until the advent of supercomputers, the magnitude of the calculations involved in transmitter-receptor interaction precluded any further understanding. Thanks to these computers, brain scientists are now able to employ the methods and theories of quantum mechanics to predict the interactions between a transmitter and its receptor. From these studies come insights that will provide the stimulus for future understanding of transmitter-receptor interactions.

For instance, in this book for the sake of simplicity we have often presented the interaction of a transmitter and its receptor as a mechanical process. Such analogies as a lock and its key are still employed even by brain scientists in discussions with one another. But analogies like this one emphasize geometry; they consider only the shapes of the transmitter and its receptor. On the whole, the molecules are treated as a system of balls and sticks that maintain their shape from moment to moment. But the chemistry of living organisms is, again, not a matter of rigid structures interacting with each other.

Analogies with rigid structures like locks and keys are even less appropriate in pharmacology, where almost all reactions are reversible. Further, most reactions involve molecules at a distance from each other, even if that distance is as short as the synaptic space between nerve cells. In a transmitter-receptor interaction, the electrons in the molecule of one exert a force, just as a magnet does, that influences the spatial distribution of the electrons in the other. Thus the interaction derives not from the transmitter influencing the receptor or vice versa but from their fields of force influencing each other. A given molecule's field of force depends upon each of its components. By analyzing the wave functions of the electrons and their energy levels (wave functions mathematically describe the coordinates of all the electrons in a molecule, together with their probable distribution in space), scientists employing differential equations can make an informed guess about the properties of the molecule and its behavior in the presence of other molecules such as receptors. Armed with this information, neuroscientists are on the way to figuring out the molecular basis for a drug's actions.

Thanks to quantum principles, it will be increasingly possible in the near future to understand drug-receptor affinity in terms of molecular configuration. Neuroscientists already know that chemicals that do not look at all alike in their chemical structure may have striking similarities in their electron distributions and therefore in their interactions with certain receptors. The "fit" of a drug with its receptor, therefore, involves dynamic and not just anatomical and geometrical considerations. This is one reason most drugs bind nonselectively—that is, not only to their specific receptors but to other sites that could not be predicted by structural considerations alone.

Multiple subtypes of receptors exist for all of the major neurotransmitters. Receptor subtypes are defined according to the potency by which they bind with various drugs. While all adrenergic (norepinephrine and epinephrine) receptors bind many of the same compounds, for example, one subtype binds epinephrine preferentially to norepinephrine, while another binds both with equal facility.

Further, since drugs often interact with more than one receptor, the final result depends on the additive effect of many different receptors activated at many different sites. Take, for instance, a drug like methamphetamine. Not only does it stimulate mental processes, it affects the heart, breathing, vision (through dilation of the pupil), and a host of other bodily processes. Such a panoply of effects results from its stimulating subtypes of the same receptor. LSD is another drug that achieves its effects by blocking subtypes of a receptor—in this case, the H-2 histamine receptor—while exerting little or no effect on another subtype, the H-1 histamine receptor. Quantum mechanical methods helped determine the critical portion of the LSD molecule that interacts with the H-2 receptor.

By combining insights from quantum mechanics with computer power, scientists should be able not only to predict what goes on when a drug interacts with a receptor but actually stimulate these interactions. They will model charge distributions, structural rearrangements resulting from drug-receptor encounters, and the likelihood that a certain drug is going to interact with a certain receptor. *Likelihood* is always the operative word when employing quantum mechanical principles. Whether a particular drug interacts with a receptor should be considered in terms of probabilities, not certainties. If the molecular "fit" is extremely close, the odds are in favor of a drug-receptor interaction. If the fit is less precise, the reaction may or may not "go," depending on statistical probabilities.

Computer modeling of a drug and its receptor will open the way for computer-graphic applications, which can be counted upon to reveal hidden structural and dynamic relations. Computer-aided design (CAD) will render much of the trial-and-error approach to drug design and testing obsolete. CAD will not simply improve on established concepts of how the brain works; it will open up new vistas, new ways of thinking about the brain. This will be particularly important in light of the enormous increase in information about the brain that is now available.

But computers cannot be expected to solve all the drug design problems looming ahead in the next two decades. According to Solomon Synder, computers have more to offer at the clinical level than at the receptor level. "Imagine a situation in which you have Valium and five other closely related drugs. A computer would be very good at saying, 'I have the feeling that if you fool around with the structure in the following way you'll get something better than the six drugs you already have.' The computer just has to look at the Valium structures. The ultimate level of computer power would involve a computer capable of evaluating the three-dimensional structure of the receptor and then have the computer tell us what kind of drug to make based on that three-dimensional structure. But that's some time off in the future because we know next to nothing about the three-dimensional structure of receptor proteins."

The practical applications of this work on receptors—brain-altering chemicals that will be available in the near future—will bring about permanent rather than merely temporary modifications in brain structure. Such drugs would exert minimal effects

on whole brain functioning, causing less sedation and tranquilization, while targeting the specific problem area. This would be
the equivalent at the chemical-molecular level of a surgeon's scal-
pel that is always directed only at abnormally functioning tissue.

Snyder says, "Soon we will be able to modify brain function in
the mentally ill in very elegant and selective ways that I can't even
predict now, but which will enable us to modulate behavior far
better than with the relatively crude drugs we have today."

But mysteries remain. "Take antagonist and agonist drugs," he
says. "Both fit into the receptor like a key into a lock. But only
the agonist causes a shape change of the receptor. The agonist
acts like the natural transmitter and stimulates the receptor to
behave as if it were being stimulated by the natural transmitter.
The antagonist, on the other hand, also occupies the receptor site
but nothing happens. It just sits there like a bump on a log and
nothing happens. Cell function doesn't change. At this point no
one can explain why and how this is so."

Moreover, while it is convenient to talk of *a* receptor, such
references are inaccurate, Snyder continues. "Every receptor ever
found has turned out to be composed of multiple receptors. The
typical story goes like this. Someone discovers a receptor. Very
soon after that, somebody else finds that the receptor is only one
of a family of receptors: A, B, C, D, et cetera. The discovery of
the marijuana receptor opens the way for the discovery of more
than one marijuana receptor. That's because we know from studying marijuana's effects that it works on different parts of the
brain. It relieves pain, improves mood, quiets anxiety, decreases
nausea, reduces the pressure within the eye in glaucoma. It's possible, indeed likely, that each of these actions is mediated by a
different receptor."

Employing technologies developed during receptor research,
neuroscientists will approach schizophrenia and manic-depressive
illnesses much as neurologists approached muscular dystrophy: at
the molecular basis, discovering the gene responsible for the illness along with the protein that the defective gene coded for. The

task will be difficult, since these diseases involve groups rather than single illnesses. Moreover, they exhibit a genetic structure much more complicated than a simple autosomal dominant or recessive character.

"The science is straightforward, and the techniques get so much better with each passing year," says Snyder. "Therefore, our estimates of how hard it will be to uncover the genetic basis for some mental illnesses changes every year. The more difficult challenge than finding or cloning a gene is finding out what the protein does that is produced by that gene. This is much harder and will be the receptor research for the near future."

Future drugs will target specific neurotransmitter-receptor interactions both at the nerve cell membrane and, by means of second messengers, inside the neuron. From this will emerge chemical maps characteristic of such conditions as introversion, extroversion, obsession, and compulsion. Already PET scan studies of the brains of people with obsessive-compulsive disorder reveal abnormalities in the basal ganglia, a group of structures under the cortex that integrate motor responses and thought processes.

Treatment of obsessive-compulsive disorder consists in giving drugs that block the reuptake of serotonin or norepinephrine or both from the synapse. Such drugs do not always eliminate symptoms completely. Rather, they are lessened or rendered less bothersome. "Since I started the new drug my symptoms have become like that tiny pebble you get stuck in your shoe. It doesn't always bother you, and sometimes doesn't bother you at all, but you never entirely forget that it's there," said one of my patients, after taking a combination norepinephrine-serotonin reuptake blocker to control his obsessive rumination about the events of his life.

In the future, lesser forms of obsession—perhaps the everyday repetitive worries that disturb our peace of mind or sleep—will be treated with a mild serotonin reuptake–blocking agent. Other behaviors with an obsessional component may yield to a similar drug approach. "Might nail-biting, for example, respond to the same therapy?" asks Dr. Judith Rapoport, chief of the child psychiatry branch of the National Institute of Mental Health. "How about other 'uncontrollable' impulse disorders such as kleptomania? Clinical trials in which such disorders are treated with antiobsessional drugs are waiting to be undertaken."

But the most interesting application of the new psychoactive agents may involve chemical attempts to modify character, personality, and habit patterns. Ordinarily these are very resistant to change, but drugs are in development that will help stimulate motivation, increase energy levels, and repair feelings of chronic low self-esteem—in short, make many people who are not suffering from a definable emotional illness feel better about themselves and the quality of their lives. As we discussed earlier, this may not be a proper use for mind- and brain-altering drugs; we are not always the ideal judges of how best to improve ourselves. But properly and sensibly used, these drugs can help us achieve the goal that many philosophical and psychological systems have suggested is our best strategy for enduring and prevailing in an uncertain world: modifying not that world but our responses to it.

✛ ✛ ✛

In the course of this book we have tracked the evolution of humankind's interest in brain- and mind-altering drugs. We started out in the primeval jungles, where for thousands of years native peoples employed sacred plants to alter their internal worlds. Along the way we encountered the early pioneers in the development of drugs aimed at the treatment of mental illness. We ended our journey within the laboratories of today's molecular chemists, where more powerful and more specific drugs are being designed that will change our brains and our internal worlds in ways that only a few years ago seemed the stuff of science fiction.

Beginning a century or so ago with an emphasis on the whole brain—the search for correlations between brain areas and human feelings and behavior—neuroscientists went on to investigate the brain on the chemical and, most recently, the molecular level.

Overall, this course is not unlike that which occurred in theoretical physics. After fifty years of searching, physicists no longer believe that the answers they seek about the universe will be found in an "ultimate particle." Instead, they emphasize trying to understand the interrelationships that hold across all levels of physical reality, from the very small to the infinitely large. This, in essence, is the unified field theory that so obsessed Albert Einstein in his later years.

A similar process is now going on within the neurosciences: the search for some fundamental organizing principle, a neurobiological unified field theory, that will unite observable human behavior with the action of molecules operating at levels far below the power of available technology to observe. Receptors provide this bridge, a bridge between the world of chemicals and the person suffering from a crippling phobia that keeps her housebound or the depressed patient who feels so much better at the end of the day than he did in the early morning when, at his worst, he even thought of suicide. Each of these situations involves an alteration in the balance of receptors and their transmitters. This mismatch between receptor and transmitter includes not only events at the chemical and molecular level but thoughts and feelings that are the correlates, at the behavioral level, of chemical changes.

As we move through the last decade of the twentieth century, which has been declared the Decade of the Brain, the unifying concept of the receptor will likely prove even more valuable as a means of advancing our knowledge of the brain and, in the process, our understanding of our thoughts, moods, and behavior. Thanks to this research, for the first time in human history we will be in a position to design our own brain.

FURTHER
READINGS

Adelman, George, ed. *Encyclopedia of Neuroscience.* 2 vols. Cambridge, Mass.: Birkhauser Boston, 1987.

Ayd, Frank J., and Barry Blackwell, eds. *Discoveries in Biological Psychiatry.* Baltimore, Md.: Ayd Medical Communications, 1984.

Black, Ira B. *Information in the Brain: A Molecular Perspective.* Cambridge, Mass.: MIT Press, 1991.

Hofmann, Albert. *LSD My Problem Child: Reflections on Sacred Drugs, Mysticism, and Science.* Los Angeles: J. P. Tarcher, 1983.

Hyman, Steven E., and Eric J. Nestler. *The Molecular Foundations of Psychiatry.* Washington, D.C.: American Psychiatric Press, Inc., 1993.

Kalivas, Peter W., and Herman H. Sampson. "The Neurobiology of Drug and Alcohol Addiction." *Annals of the New York Academy of Sciences,* vol. 654 (1992).

Lewin, Louis. *Phantastica: Narcotic and Stimulating Drugs—Their Use and Abuse.* New York: E. P. Dutton & Company, 1931.

Matossian, Mary Kilbourne. *Poisons of the Past: Molds, Epidemics, and History.* New Haven, Conn.: Yale University Press, 1989.

Ray, Oakley S. *Drugs, Society, and Human Behavior.* St. Louis, Mo.: The C. V. Mosby Company, 1972. Second edition, with Charles Ksir, 1990.

Schultes, Richard Evans, and Albert Hofmann. *Plants of the Gods.* New York: Alfred van der Marck Editions, 1987.

Schultes, Richard Evans, and Robert F. Raffauf. *Vine of the Soul: Medicine Men, Their Plants and Rituals in the Colombian Amazonia.* Oracle, Ariz.: Synergistic Press, 1992.

Smith, C. U. M. *Elements of Molecular Neurobiology.* New York: John Wiley & Sons, 1989.

INDEX

Aberdeen, University of, 173
Acetylcholine, 23, 24–25, 37, 116–18, 132–33, 205, 207
Acetylcholine receptor blockers, 38–39
Aconite, 36
Action potential, 19
Addiction, 10–11, 107–8, 145–52
  and access to drugs, 150–51
  and behavior, 178
  and brain abnormalities, 147, 148–49
  changes in attitude about, 10
  genetic basis of, 150
  neurological basis for, 178
  and nucleus accumbens, 177
  and personality, 150
  physical vs. psychological, 149
  and positive vs. negative reinforcement, 151
  and self-medication, 152
  See also specific substances.
Adenosine, 124
Adenylate cyclase, 187–88
Adrenaline, 207
Affinity, 83

Age-associated memory impairment (AAMI), 196
Aging
  and forgetting, 196
  and Parkinson's disease, 114
Agonist vs. antagonist drugs, 170, 213
Air Surgeons Bulletin, 138
Alcohol, 45
  and anxiety, 157
  and barbiturates, 154
  and dopamine, 145–46
  and G protein function, 204
Alkaloids, 37, 38
Alzheimer's disease, 25
  and excitotoxic assault, 132, 133
  and memory, 132, 195
  and nicotine, 120–21, 195
American Psychiatric Association, 120
Amino acids, 23, 94
  and diversity, 99
  excitatory, 129, 130–34
  multiplicity of, 95–96
  sequence of, 96
AMP (adenosine monophosphate), 188

Amphetamine, 17, 137–41, 144–46, 176–77
    as "diet pill," 201
    effects of, 77, 125, 137, 138–40, 211
    "ice," 140–41
    "speed," 45, 138, 140
Analeptic, 47
Analogues, synthetic, 208
Anandamide, 190–91
Anatomy of Melancholy, The (Burton), 76
Angel dust. See PCP.
Angiotensin, 206
Anhedonia, 146
Anorexia, 201
Antagonists, 83, 170, 213
Antianxiety agents, 158–67
Anticholinergic agents, 37
Anticonvulsants, 153–54
Antidepressants, 81–84, 152
    and sexual disorders, 197
    tricyclic, 82–83
Antihistamines, 69, 70
Antipsychotic drugs, 72
Anxiety, 155–62
    antianxiety agents, 158–67
    biologic substrate of, 159
    vs. fear, 156
    as signal, 161
Anxiety neurosis, 124
Aphasia, 90–91
Aspartate, 130
Asthma, 135–37
Ataraxia, 69–70
Atropine coma therapy, 39
Autonomic nervous system, 16–18, 69, 71, 84, 135–36
Autoradiography, 110, 117, 174–75
Autosomal dominant pattern, 100
Ayahuasca, 33–34, 35
Aztecs, 51–53

Baby food, and MSG, 133
Balzac, Honoré de, 123
Banded krait, prey paralyzed by, 117
Barbital, 153
Barbiturates, 8, 153–54, 157
Beckett, Arnold, 170
Behavior
    and addiction, 178
    and function, 95
Behavioral level, 30, 83, 205
Bein, Hugo, 75
Belladonna, 36, 38, 40

Benzedrine, 137–38, 140
Benzodiazepines, 151, 164–66
Benzophenones, 163
Berger, F. M., 158–61
Bergland, Richard, 207
Bernhardt, Sarah, 143
Bhang, 184
Binding sites, 175, 211
Bini, Lucio, 78
Biological link, need for, 84
Biological treatments, 8
Biphasic curve, 118
Bipolar disorder, 61–67
Blood-brain barrier, 138
Blood pressure, 136
Blood sugar, and brain injury, 132
Blundell, Tom, 194
Body, vs. mind, 9
Bohr, Niels, 209
Bonner, Tom, 188
Boston University School of Medicine, 177
Bradley, William, 158
Braestrup, C., 165
Brain
    biochemical changes in, 59–60
    cells in, 98
    chemical alteration of. See Drugs; Mind-altering substances.
    on chemical and molecular level, 6, 60, 84
    chemical dialogue in, 209
    chemical diversity in, 26
    chemical transmissions in, 205
    function, new model of, 204–6, 207
    functions of, 5–6, 12–32
    genetic factors in, 99
    as gland, 207
    and information, 7–8, 19, 89–96, 101
    injury to, 24, 132
    levels of, 30, 91–93
    mapping of, 202
    memory structure of, 31
    in mental dysfunction, 9, 72, 76
    and mind, 59–60, 73
    nature of, 206–7
    organization of, 12–32
    and research difficulties, 193–94
    self-destruction of, 133
    as self-reflecting organ, 28–29, 30
Brain cells, forms of, 98
Brain hedonism, 146
"Breathing," of protein molecule, 209

British Drug Houses Ltd., 158
Broca's aphasia, 90–91
Bromides, 8
Burton, Robert, 76
BZ receptors, 166–67

Caapi, 34
Cade, John, 61–67, 84, 162
Caffeine, 45, 122–25
Calcium transport, 131
Cameron, Alister, 69
Camphor oil, 77–78
Cancer, 122
Cannabinols, 35
Cannabis. *See* Marijuana.
Carrillo, George, 108
Cascade reaction, 187–88, 203–4
Catatonic schizophrenia, 103
Catecholamine system, 141
Cattell, Raymond, 156
Cell
  brain, forms of, 98
  death, 101, 114–15
  DNA in nucleus of, 18
Cerebral cortex, 139–40
Cerletti, Ugo, 77–78
Charas, 184
Chemical linkages, 206
Chemical maps, 214
Chemical processes
  and association of mind and brain, 59–60, 73
  and mental illness, 9, 59–60, 77
  and mind alteration, 9–10, 46, 178
Chemicals, regulatory, 207–8
Chemical substances
  and benefit vs. damage, 39–40, 115
  food additives, 133
  radioactive tagging of, 110–11, 117, 165, 172, 174
Chemistry, theoretical, 163
Chloral hydrate, 153
Chlordiazepoxide, 164
Chloroform, 45
Chlorpromazine, 71–72, 73, 75–76, 81, 84, 160, 162
Cholinergic antagonists, 39, 133
Chromosome eleven, 208
Chromosome four, 100–101, 208
Ciba Laboratories, 75
Coca-Cola, 143
Cocaine, 10, 44, 121, 142–52, 176–78
  crack, 146–47, 148, 150

Freud's uses of, 142–43
Codeine, 44
Coffee, 45, 122–25
Cold remedies, over-the-counter, 39
Collins, Alan, 118
Colorado, University of, 118
Communication
  and decoding by brain, 89–94
  distortions of, 103
  by neurons, 4–5, 18–22, 91–92
  neurotransmitters in, 21
  parasynaptic, 205–6
Competition, in research, 162, 174
Computer-aided design (CAD), 212
Computers, in research, 191, 209–10, 211–12
Conan Doyle, Sir Arthur, 144
Connecticut Medical Society, 144
Control, by drugs, 8, 11, 32
Crack cocaine, 146–47, 148, 150
Creatinine, 62
Crick, Francis, 98
Cyperquat, 111–12

Δ-9-THC (Δ-9-tetrahydrocannabinol), 183–84, 190
Darwin, Charles, 98
Decade of the Brain, 216
Delay, Jean, 72
Demerol, 105
Dendrite, and nicotine, 116
Deniker, Jean, 72
Deoxyribonucleic acid. *See* DNA.
Dependence. *See* Addiction.
Deprenyl, 113, 114
Depression, 74–85
  and alcohol, 157
  and barbiturates, 157
  biological link needed for, 84
  and catecholamine system, 141
  chromosome-associated, 208
  and dopamine, 148
  endogenous vs. exogenous, 76–77, 82
  functional neuroanatomic circuit in, 195
  and G protein function, 204
  and MAO-A, 112
  objective vs. subjective components of, 82
  origin of, 80–81
  treatments for, 78–81
  and vital disturbances, 82
  See also Mental illness.
De Ropp, Robert S., 40, 59

Descartes, 5
Design cycle, receptor-based, 193–94
Devane, William, 190
Diagnostic and Statistical Manual (DSM-III-R), 120
Diazepam, 164–65
"Diet pills," 201
DiMaggio, Joe, 156–57
DNA
    in cell nucleus, 18
    and Huntington's disease, 100, 101
    information transferred from, 96–97
    and internal cellular processes, 178–79
    molecular arrangements within, 91
    recombinant, 208
    and RNA, 97–98, 101
    structure of, 94–98
Domoic acid, 132
Dopamine, 24, 84, 94
    and amphetamine, 141
    and depression, 148
    functions, 110–11
    and "high," 145–46, 175
    and Parkinson's disease, 105, 109–12
    and reward systems, 119
    structure of, 138
    and tardive dyskinesia, 198
Dosage, and poison vs. medicine, 39–40, 115
Dostoevsky, Fyodor, 55
Drugs
    and addiction. *See* Addiction.
    antiobsessional, 215
    binding of, 211
    classification of, 43
    in context, 50
    control by, 8, 11, 32
    cost of, 192
    design of, 191–201, 208–9
    functions of, 161
    good vs. bad, 121–22, 129
    interaction with receptors, 192–93, 211
    intrinsic (primary) reinforcing powers of, 146
    market demand for, 199–200
    mind-altering. *See* Mind-altering substances
    multiple effects of, 43
    and Parkinson's disease, 105–15
    power of, 42–43
    reasons for taking, 43
    smoking of, 146–47, 185
    and toxic equation, 43

    See also Mind-altering substances.
Drug-seeking behavior, 176
Drugs and the Mind (De Ropp), 40, 59

Eating disorders, 201
Eco, Umberto, 7
ECT (electroconvulsive therapy), 8–9, 77–78
Edison, Thomas, 143
Effectors, enzyme, 203
Ehrlich, Paul, 22
Electric eels, 117
Electric shock treatment, 8–9, 77–78
Electrons, spatial distribution of, 210
Endogenous depression, 76–77, 82
Endorphins, 24, 119
    enkephalin, 174
    as natural opiates, 148–49, 172–73
    and personality, 175
Energy-enhancing drugs, 199–200
Enkephalin, 174
Environment, and genes, 98
Enzymes, 23, 94
    effectors, 203
Ephedrine, 135, 136, 138
Epileptic seizures, 77, 153–54, 166
Epinephrine, 58
Ergot, 46–49, 57
Erythroxylon coca, 142
Ether, 45
Ethics, 199
Etorphin, 170–71
Europhiants, 43–44
Excitants, 45
Excitotoxins, 131–34
Exogenous depression, 76–77, 82
Exorcisms, 46
Experience, and heredity, 98
Eye, neuroactive agents in, 27–28

Fabing, Howard, 69
Fabric of the Mind, The (Bergland), 207
Fatigue, as signal, 200
Fear, vs. neurotic anxiety, 156
Feedback loop, and fatigue, 200
Flexibility, waxy, 104
4560 RP, 70–71
Freud, Sigmund, 76, 142–43, 156
Function, and behavior, 95

Galen, 135–36
Gamma-amino butyric acid (GABA), 23, 166

Gangrene, 47
Ganja, 184
Gene probe, 189
Genes
  and addiction, 150, 178–79
  altering structure of, 150, 208, 214
  and environment, 98
  talking, 102
Genetic defects, corrective measures for,
      150, 213–14
Georgetown University, 117–18
Gilman, Alfred G., 204
Glaxo, 196
Glutamate, 130–34
G proteins, 203–4
G22355 (imipramine), 81–82, 85
Guinea pig ileum, 173
Gusella, James, 100
"Gut" feelings, 206

Hallucinogens, 33–41
  chemical synthesis of, 55–56, 57
  effects of, 35, 37–39, 47–55, 57, 58
  and neurotransmitters, 34–37, 58–59
  phantasticants, 44–45
  work within brain, 46
Halsted, W. S., 142
Harmine, 34
Harrison Narcotics Act, 144
Harvard Medical School, 148
Hashish, 184
Hebrew University, 190
Heffter, Arthur, 45, 52
Heim, Roger, 53–54
Helmreich, Ernst J. M., 204
Hemispheres, of brain, 12–13
Hemlock, 36
Hemp plant, 183
Henbane, 36, 39–40
Heredity, 96–98, 101
Herkenham, Miles, 186, 187, 189, 204–8
Hernandez, Francisco, 57
Heroic Age, of research, 162
Heroin, 10, 44, 146, 149
  and cancer, 122
  home-brewed substitutes, 105–8, 110
Hexing herbs, 37, 38
High blood pressure, 78
Hippocrates, 65, 76
Histamine, and shock, 69
Hoffmann-LaRoche laboratories, 163, 164
Hofmann, Albert, 35, 37, 46–60, 72–73,
      84

Holbein, Hans, 35
Homer, 39
Hormones, regulatory, 207
Howlett, Allyn, 187–88, 190
Huautla de Jiménez, 53
Hughes, John, 173–74
Humors, and mental illness, 65, 76
Huntington's chorea, 99–102, 133, 208
Hyperactivity, 152, 196–97, 198–99
Hypnotics, 45, 153
Hypodermic syringe, invention of, 142
Hypothalamus, 140

Imipramine, 81–82, 85
Imperial Cancer Research Fund Unit of
      Structural Molecular Biology, 194
Impotence, 197
Indian Medical World, 74–75
Inebriants, 45
Information
  and brain, 7–8, 19, 89–96, 101
  flow of, 98, 205–6
Informational disease, 94–95
Information levels, 91–93
Information processing, and cascades, 203
Information transfer, 91
Insulin coma therapy, 8
Intoxication, inebriants, 45
Intoxication: Life in Pursuit of Artificial
      Paradise (Siegel), 60
Intuition, in research, 162
Ions, distribution of, 19, 21
Isoniazid, 78–80
Isoniazid-reserpine treatment, 79

Johns Hopkins University School of
      Medicine, 171, 202
Johnson, M. Ross, 186
Jones, John, 170

Kellar, Kenneth, 118, 121
Khantzian, Edward, 148
King, Roy, 148
Kline, Nathan, 79, 162
Kornetsky, Conan, 177
Kosterlitz, Hans, 173–74
Kuhn, Roland, 81–82, 84, 162

Laborit, Henri-Marie, 68–72, 75, 84
Lange, J., 81
Langston, William, 108
Language disturbance, 90–91
LA-111, 58

Laudanum, 169
L-dopa, 105, 106–10, 113–14
Leary, Timothy, 60
Levitation, 36, 38
Levonantradol, 186–88
Lewin, Louis, 42–46, 52, 184
Librium, 164–65
Limbic system, 13, 16
  and emotions, 93, 157, 172
  and hallucinogens, 55
  and nucleus accumbens, 177
  and pleasure response, 175
  stimulation of, 140
Linder, Maurine E., 204
Lithium, 61–67, 73, 84, 204
Locke, John, 155
Locus ceruleus, 139
Loewi, Otto, 24–25
LSD (lysergic acid), 35, 46–55, 57–60, 84, 211
Ludwig, B. J., 159
Lysergic acid amide, 57–58

Ma huang, 135
Mandrake, 35
Manic-depressive illness. See Mental illness.
MAO (monoamine oxidase), 79–80
MAO-A, 112
MAO-B, 112, 113–14
MAOIs (MAO inhibitors), 80, 84–85
Maps, and information levels, 92–93
Mariani, Angelo, 143
Marijuana, 10, 35, 112, 145–46, 183–86
  chemical diversity of, 183–84
  chemical-molecular mode of action of, 187
  medical use of, 184–85, 191
Marijuana gene, discovery of, 190, 206
Marijuana receptor, 183–201, 213
  cloning of, 188–90
  mapping of, 187, 189
Markey, Sanford P., 109, 114
Massachusetts General Hospital, 100
Mass hysteria, 38
Matsuda, Lisa, 188–90, 206
Mechoulam, Raphael, 190
Medical Journal of Australia, The, 64
Medicine
  vs. poison, 39–40, 115
  and user's forgetfulness, 64
Meduna, Ludwig von, 77
Melancholy. See Depression.

Melvin, Lawrence, 186
Memory
  and Alzheimer's disease, 132, 195
  and forgetting, 196
  and hallucinogens, 37–39
  and harmful chemicals, 133
  and regulatory chemicals, 207–8
  structure of, 31
Memory pills, 196
Mental illness
  biochemical causes of, 51
  biological treatments for, 72
  and bodily fluids, 65, 76
  and brain functioning, 9, 72, 76
  and chemical processes, 9, 59–60, 77
  and molecular genetics, 213–14
  and neurotransmitters, 80–81, 83–84
  and toxins, 61–67
  treatments, 8–9, 196
  See also specific illnesses.
Meperidine, 105, 195
Mephenesin, 158–59
Meprobomate, 159–63
Mescaline, 34–35, 44–45, 52, 58
Messages
  of brain, 4–5, 18–19
  decoded and encoded, 89–96
  distortions, 103
Metaphors
  for brain function, 5–6
  of magic circle, 58
Methamphetamine. See Amphetamine.
Methylphenidate, 45, 152
Meynert, nucleus basalis, 25, 132–33
Mickey Finn, 153
Micromolecules, 7
Microscopic level, 30
Miltown, 159
Mind
  vs. body, 9
  and brain, 59–60, 73
Mind-altering substances
  and biological action, 83, 186
  categories of, 43–46
  and chemical processes, 9–10, 46, 178
  chemical synthesis of, 55–56, 57
  ethical questions about, 199–200
  future research on, 186–95, 211–15
  history of, 33–41, 215–16
  permanent harm from, 177, 198–99
  and physical functions, 46
  from plants, 10–11, 41, 42–60
  as time bomb, 198–99

unity of action among, 176–77
See also Drugs.
"Mismatches" between transmitter and
    receptor, 204–5
Molecular genetics, 99, 213–14
Molecular level, 30, 91, 92–93, 178
Molecule, disordered, 31–32, 93, 101
Molecules, distance between, 210
Monkshood, 36
Monoamines, 23, 79–80, 112
Morphine
    and amphetamine, 176–77
    and cancer, 122
    and cocaine, 142–43
    as pain reliever, 169–70
Motor cortex, and nucleus accumbens,
    177
Mourning and Melancholia (Freud), 76
MPPP, 105–8
MPP+, 111–14
MPPP-MPTP, effects of, 107
MPTP, 106–8, 110–12, 113–15
MSG (monosodium glutamate), 133–34
Mushrooms, hallucinogenic, 53–56, 58–59
Mussel poisoning, 132
Mysteries of Opium Revealed, The (Jones),
    170

Nalorphine, 170
Naloxone, 170–71, 173, 176–77
Narcolepsy, 136
National Institute of Diabetes and
    Digestive and Kidney Disorders
    (NIDDK), 187
National Institute of Mental Health
    (NIMH), 106–9, 111, 186, 188–89,
    204, 215
National Institute on Drug Abuse, 186
"Natural highs," 141
Nerve cell membrane
    and cocaine, 142
    and neurotransmitter receptor, 203–4,
    214
Networks, of neurons, 91–92
Neuroanatomic circuit, 195
Neuroendocrine activity, 133
Neuroleptics, 72, 198
Neuromedin, 188
Neuromelatonin, and MPP+ toxicity, 113
Neuromodulation, 28
Neurons
    components of, 18
    firing of, 21–22, 92, 130

and glutamate, 130
and good vs. bad drugs, 121
and information levels, 91, 92–93
messages sent by, 4–5, 18–22, 91–92
networks of, 91–92
nucleic acids in, 97
structure of, 96
Neuropeptides, 23–24, 25–26
Neurophysiology, 123
Neuropsychiatry, 8–9
Neuroses, treatments, 8
Neurotransmitters
    biochemical understanding of, 195
    and cascade reaction, 188
    chemical types of, 23–25
    in communication, 21
    function of, 27–29
    and hallucinogens, 34–37, 58–59
    and Huntington's disease, 101
    inhibitory, 124
    and mental illness, 80–81, 83–84
    and receptors, 31, 203–5, 209–10, 214
    and reuptake, 81–83
    and second messengers, 203–4
    structures of, 22
New York Times, The, 148
Nicotine, 116–21, 145–46
    and Alzheimer's disease, 120–21, 195
    and negative reinforcement, 151
Nicotine receptors, 117–18, 195–96
Nightshade, deadly, 35–36
Nitrogen, 34
NMDA (N-methyl-D-aspartic acid), 130–
    32
Noradrenaline, 207
Norepinephrine, 17–18, 24, 34–35, 83, 94,
    138
    and amphetamine, 141
    and caffeine, 123
    and MAOIs, 80
    and mescaline, 34, 58
    and peptides, 205
*Notes of a Botanist on the Amazon and
    Andes* (Spruce), 34
Nucleic acids, 96–98
Nucleus accumbens, 145, 177, 178
Nucleus basalis, 25, 132–33

Obsession, 214–15
Olney, John, 133
Ololiuqui, 57–60
On the Origin of Species (Darwin), 98
Opiate receptors, 170–75

Opiates, 168–78
   and dopamine, 145–46
   and G protein function, 204
   and naloxone, 176–77
   natural (endorphins), 148–49, 172–73
   as pain relievers, 169–70, 172, 175
   synthesized, 174
Opium, 44, 81, 168–69
Oxazepam, 165
Oxymorphone, 170

Pacing, marijuana intake, 185
Pain, and substantia gelatinosa, 174–75
Palsy, shaking, 104
Panic attack, 124, 200
Panic reactions, to marijuana, 185
Paraldehyde, 153
Paranoid schizophrenia, 137
Paraquat, 111–12, 113
Parasympathetic nervous system, 37, 136
Parasynaptic communication, 205–6
Parke-Davis, 126
Parkinson, James, 104–5
Parkinson's disease, 24, 72, 104–5
   as consequence of aging, 114
   drug-induced, 105–15
   and excitotoxins, 133
   and substantia nigra, 114, 140
   treatments, 113–14
Patent remedies
   cocaine in, 143–44
   opium in, 169
PCP (phencyclidine), 126–34
   as animal anesthetic, 126–27
   and brain damage after stroke, 131
   as easily synthesized, 127
Peirce, C. S., 31
Penicillin, 158
Peptides, 25–26, 205–8
Peroxide, 113
Person, patient as, 102
Personality
   addictive, 150
   and drugs' appeal, 43
   and endorphin systems, 175
Pert, Candace, 171–72, 174
Peterson, Robert C., 186
PET (positron emission tomography) scans, 110–11, 174, 214
Peyote, 34–35, 42, 44–45, 52
Pfizer Inc., 186

Phantastica: Narcotic and Stimulating Drugs (Lewin), 42–46
Phantasticants, 44–45
Pharmacology, 210
Phencyclidine. See PCP.
Phenobarbital, 153
Phenothiazines, 162, 163
Phenylalanine, 94–95
Phenylethylamine, 35
Phenylketonuria (PKU), 94–95
Phobia, 157
Picoseconds, defined, 209
Pig brain extract, 173, 174
Plants
   compounds derived from, 191
   mind-altering ingredients of, 10–11, 41, 42–60
   sacred, 52–60
Plants of the Gods (Schultes and Hofmann), 35, 56
Pleasure response, 140, 145, 146, 175–76, 178
Pliny, 40
Poisons
   action of, 22
   alkaloids, 37
   vs. medicines, 39–40
Protein molecules, 209
Proteins, 94
   and diversity, 99
   and heredity, 96–98, 101
   and thermal energy, 209
Prozac, 200, 202
Psilocin, 55, 58–59
Psilocybin, 55–56, 57, 58–59
Psychedelic movement, 60
Psychiatric Research, 107
Psychic disturbances, biochemical causes of, 51
Psychic energizers, 79
Psychoactive agents, 215
Psychomotor retardation, 78
Psychopathology, 199–200
Psychopharmacology, 67, 84
   and drug design, 192–93, 208–9
   and good vs. bad drugs, 121–22
Psychosis
   causes of, 132
   and chlorpromazine, 72
   PCP-induced, 126, 132
   and reserpine, 75
Psychotherapy, 78, 127
Pure Food and Drug Act, 144

Quantum mechanics, 209–11

Radioactive tagging, of chemical
    substances, 110–11, 117, 165, 172,
    174–75
Randall, Lowell O., 164
Rapoport, Judith, 215
Rat poison, 108
Rauwolfia, 74–75, 78
Receptor autoradiography, 174–75
Receptor-based design cycle, 193–94
Receptors
    and affinity, 83
    binding techniques, 174–75
    biochemical understanding of, 195
    and brain self-relatedness, 30–31
    as bridge, 216
    drug interaction with, 192–93, 211
    multiple subtypes of, 211, 213
    and neurotransmitters, 31, 203–5, 209–
        10, 214
Reinforcement, 146, 151
Relaxation, 152–54
Religious ceremonies, sacred plants in, 52–
    56
Renin, 193, 206
Repetition, in communication, 91
Research
    brain, difficulties of, 193–94
    competition in, 162, 174
    computers in, 191, 209–10, 211–12
    future, 186–95, 208–15
    and safety, 194–95
    by teams vs. individuals, 162–63
    technologies, 213–14
Reserpine, 75–76, 78, 79–80
"Restless legs" syndrome, 196–97
Reuptake process, 80–83, 85, 141
Reverse tolerance, 185
Rhône-Poulenc, 70–71
Rice, Kenner, 187
Richardson, Benjamin, 153
Ritalin, 45, 152, 198–99
RNA (ribonucleic acid), 97–98, 101
Ro-5-0690, 164
Rolling Stones, 164
Rush, Benjamin, 70

Sabat (Black Mass), 36, 38, 39–40
Sabina, Maria, 54–55, 56
Safety, and research, 194–95
St. Louis University Medical School, 187
Salem witch trials, 46, 47, 100

Sandoz, 46, 47, 53
Schizophrenia, 51, 72, 77, 132, 213
    catatonic, 103
    and G protein function, 204
    paranoid, 137
    and PCP symptoms, 126, 127–29
Schultes, Richard Evans, 35, 37, 56
Schwartz, Rochelle, 118
Scientific American, 66, 204
Scopolamine, 38–39
Scott, Charles, 79
Second messengers, 203–4
Sedation, 152–54
    preoperative, 70–71
Seizures, and brain injury, 132
Selegiline, 113
Self-medication, and addiction, 152
Sense organs, 26–29
Sensory isolation, from PCP, 128
Serax, 165
Sernyl, 126
Serotonin, 24, 55, 59, 83, 205, 207
Serotonin reuptake blockers, 197, 202,
    214–15
Serturner, Friedrich Wilhelm, 169
Sexual disorders, 197
Shamans, 51, 54–55
Sherrington, Sir Charles, 5
Shock, use of term, 68–69
Siegel, Ronald, 60
Sinsemilla, 184
Skates, electric, 117
Skin patch, transdermal, 120
Sleeping pills, 45, 153
Smokers
    nicotinic receptors of, 118
    withdrawal symptoms of, 119–20
Smoking, as efficient drug delivery system,
    146–47, 185
Snakes, prey paralyzed by, 117
Snyder, Solomon, 171–72, 174, 202, 212–
    13, 214
Social custom, and neurophysiology, 123
Social withdrawal, 200
Société Médico-Psychologique, 71
Society
    considerations of, 199
    neuroticism of, 201
Somatic treatments, 8
Soot, 36
Spath, Ernst, 45
"Speed." See Amphetamine.
Spruce, Robert, 34

Squires, Richard F., 165, 167
Stanford Medical School, 148
Sternbach, Leo H., 163–64
Stimulants. *See* Excitants.
Stoics, 69
Stroke
    and excitotoxic damage, 131, 132, 134
    and language disturbance, 90
Structure, and function, 95
Substance K, 24, 188–89
Substance P, 24, 188, 206
Substantia gelatinosa, 174–75
Substantia nigra, 24, 72, 84
    cell loss in, 114–15
    functions of, 109–10, 140
    and MPP+, 111–13
    and Parkinson's disease, 114, 140
Supercomputers, 209–10
Sydenham, Sir Thomas, 169
Symbols and signs, 31
Sympathetic nervous system, stimulation
    of, 135, 136
Sympathomimetic drugs, 17
Synapses, and information levels, 91, 92–
    93
Systems, and information levels, 92–93

Tardive dyskinesia, 197–98
Temporal lobe epileptics, 55
Teonanacatl, 52–56, 58–59
Terenius, Lars, 175
Texas, University of, Southwestern
    Medical Center, 204
THC, 183–84, 190
Theobromine, 123
Theophylline, 123
Thermal energy, and proteins, 209
Thorazine, 198
Tobacco. *See* Nicotine.

Torticollis, 72
Tosteson, Daniel C., 66
Toxic equation, 43
Toxins, and bipolar disorder, 61–67
Tranquilizer guns, 127
Tranquilizers, 70, 84
    and anxiety research, 159–67
    proper use of, 160–61
    and tardive dyskinesia, 198
Trauma, and brain injury, 132
Tricyclic antidepressants, 82–83
Tryptamine, 35, 55, 59
Tuberculosis, treatments, 78–79
Tukano Indians, 34

Unified field theory, 216
Unimed, 184–85
University of Uppsala, 175
Urea, 62–63
Uric acid, 62, 63, 66

Valium, 151, 164–65, 167
Vegetable protein, hydrolized, 133
Ventral tegmentum, 145, 178
Verne, Jules, 143
Vin Mariani, 143
Vital disturbances, 82

Warner Laboratories, 79
Washington University, 133
Wasson, R. Gordon, 53
Wernicke's aphasia, 90
Witchcraft, 35–38, 40, 46, 47
Withdrawal reaction, 176
Word salad, and stroke, 90
Workaholism, 199–200
Würzburg Medical School, 204

Xanthines, 122–24